Springer Theses

Recognizing Outstanding Ph.D. Research

For further volumes:
http://www.springer.com/series/8790

Aims and Scope

The series "Springer Theses" brings together a selection of the very best Ph.D. theses from around the world and across the physical sciences. Nominated and endorsed by two recognized specialists, each published volume has been selected for its scientific excellence and the high impact of its contents for the pertinent field of research. For greater accessibility to non-specialists, the published versions include an extended introduction, as well as a foreword by the student's supervisor explaining the special relevance of the work for the field. As a whole, the series will provide a valuable resource both for newcomers to the research fields described, and for other scientists seeking detailed background information on special questions. Finally, it provides an accredited documentation of the valuable contributions made by today's younger generation of scientists.

Theses are accepted into the series by invited nomination only and must fulfill all of the following criteria

- They must be written in good English.
- The topic should fall within the confines of Chemistry, Physics, Earth Sciences, Engineering and related interdisciplinary fields such as Materials, Nanoscience, Chemical Engineering, Complex Systems and Biophysics.
- The work reported in the thesis must represent a significant scientific advance.
- If the thesis includes previously published material, permission to reproduce this must be gained from the respective copyright holder.
- They must have been examined and passed during the 12 months prior to nomination.
- Each thesis should include a foreword by the supervisor outlining the significance of its content.
- The theses should have a clearly defined structure including an introduction accessible to scientists not expert in that particular field.

Tian Lu

Nanomaterials for Liquid Chromatography and Laser Desorption/Ionization Mass Spectrometry

Doctoral Thesis accepted by
The Ohio State University, Columbus, USA

Author
Dr. Tian Lu
Analytical Department
Ashland Inc.
Bridgewater, NJ
USA

Supervisor
Prof. Susan V. Olesik
Department of Chemistry and Biochemistry
The Ohio State University
Columbus
USA

ISSN 2190-5053 ISSN 2190-5061 (electronic)
ISBN 978-3-319-36175-8 ISBN 978-3-319-07749-9 (eBook)
DOI 10.1007/978-3-319-07749-9
Springer Cham Heidelberg New York Dordrecht London

Softcover reprint of the hardcover 1st edition 2015

Printed on acid-free paper

Springer is part of Springer Science+Business Media (www.springer.com)

Parts of this thesis have been published in the following journal articles:

Lu, T.; Olesik, S. V. "Homogeneous Edge-Plane Carbon as Stationary Phase for Reversed-Phase Liquid Chromatography" *Anal. Chem.* **2015**, DOI: 10.1021/ac503195r.

Lu, T.; Zewe, J. W.; Cevheri, N.; Olesik, S. V. "Fluorescent Carbon by Covalently Attaching a BODIPY Fluorophore" *RSC Adv.* **2014**, 4, 62651.

Lu, T.; Olesik, S. V. "Electrospun Nanofibers as Substrates for Surface-Assisted Laser Desorption/Ionization and Matrix-Enhanced Surface-Assisted Laser Desorption/Ionization Mass Spectrometry" *Anal. Chem.* **2013**, 85, 4384.

Lu, T.; Olesik, S. V. "Electrospun Polyvinyl Alcohol Ultra-Thin Layer Chromatography of Amino Acids" *J. Chromatogr., B* **2013**, 912, 98.

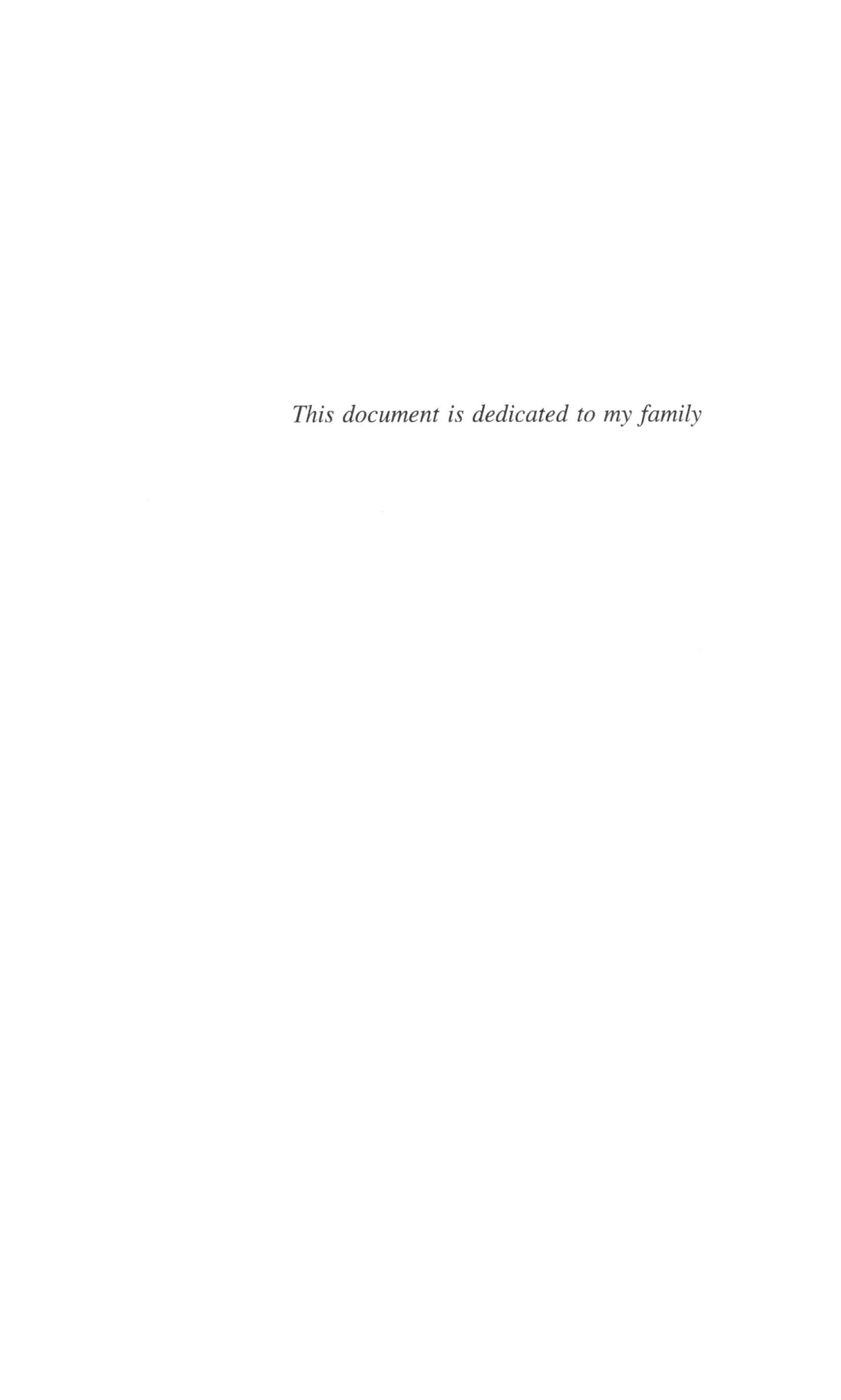

This document is dedicated to my family

Supervisor's Foreword

I would like to acknowledge the significant innovations of Tian Lu that are highlighted within this dissertation.

This dissertation is broad sweeping in its scope spanning from chemical to synthesis to state-of-the-art analytical chemistry. Major improvements in ultrathin layer chromatography, surface-assisted laser desorption ionization (SALDI), and matrix-assisted (ME-SALDI) mass spectrometry. In addition, an edge-plane-based, ordered carbon surface is highlighted that provides unique selectivity in liquid chromatography. Electrospun polyvinyl alcohol (PVA) ultrathin layer chromatographic (UTLC) plates were fabricated using in situ crosslinking electrospinning technique. The efficiency on PVA plates was greatly improved compared to the efficiency on silica HPTLC plates.

Electrospun polymer nanofibers (polyacrylonitrile, polyvinyl alcohol, and SU-8 photoresist), and carbon nanofibers were used for the first time as substrates for surface-assisted laser desorption/ionization (SALDI) and matrix enhanced surface-assisted laser desorption/ionization (ME-SALDI) analyses. Mass spectra of organic polymers with high signal/noise were observed. For example, polyethylene glycol with a molecular weight as high as 900,000 Da was successfully detected using the carbon nanofibrous substrate, which is the highest molecular weight that has been studied by SALDI to date. High quality polystyrene mass spectra were obtained for the first time using SALDI nanofibrous polyacrylonitrile substrates. ME-SALDI showed enhanced signal for the analytes using electrospun nanofibrous substrates compared to using a conventional stainless steel substrate. The detection limit of 400 amol was achieved for angiotensin I using the carbon nanofibrous ME-SALDI substrate.

This is exquisite and groundbreaking work; I congratulate Dr. Lu on the success that she has had to date.

Columbus, USA, May 2014 — Prof. Susan V. Olesik

Acknowledgments

First, I would like to express my great thanks to my advisor, Dr. Susan Olesik, for her continuous support for my Ph.D. program. Her inspiring guidance helped me to make progress during my Ph.D. research. She also encouraged my own ideas and different ways to solve problems.

I would like to extend my special thanks to Dr. Joe Zewe who started the fluorescent carbon synthesis and worked closely with me on the project. He also helped me on the optimization of the solid phase extraction condition. I would like to thank Prof. Minami Yoda and her group members for the suggestions on the synthesis of fluorescent carbon particles and providing the image of the particles. I also thank the National Science Foundation for providing the funding to complete this work.

I would like to thank all of the Olesik group members who gave me insightful suggestions on my research and it was a pleasure to work with them.

Last but not least, I would like to thank my family members. Without their support and encouragement this dissertation would not have been possible.

Contents

Abbreviations

ACN	Acetonitrile
AHAMTES	N-(6-Aminohexyl)aminomethyl triethoxysilane
BODIPY	4,4-Difluoro-5,7-dimethyl-4-bora-3a,4a-diaza-*s*-indacene
CHCA	α-Cyano-4-hydroxycinnamic acid
CVD	Chemical vapor deposition
DART	Direct analysis in real time
DHB	2,5-Dihydroxybenzoic acid
DMF	N,N-Dimethylformamide
DR	Dithranol
EFLC	Enhanced-fluidity liquid chromatography
FITC	Fluorescein isothiocyanate
FWHM	Full width at half maximum
GA	Glutaraldehyde
GLAD	Glancing angle deposition
H	Theoretical plate height
HILIC	Hydrophilic interaction liquid chromatography
HOPG	Highly ordered pyrolytic graphite
HPTLC	High performance thin layer chromatography
hR_F	Retardation factor
k	Retention factor
LC	Liquid chromatography
LSER	Linear solvation energy relationship
MALDI	Matrix-assisted laser desorption/ionization
ME-SALDI	Matrix-enhanced surface-assisted laser desorption/ionization
MS	Mass spectrometry
N	Theoretical plate number
PALDI	Polymer-assisted laser desorption/ionization
PAN	Polyacrylonitrile
PEG	Poly(ethylene glycol)
PS	Polystyrene
PVA	Polyvinyl alcohol
S/N	Signal to noise ratio
SALDI	Surface-assisted laser desorption/ionization
SELDI	Surface-enhanced laser desorption/ionization

SEM	Scanning electron microscope
SNAP	Sophisticated numerical annotation procedure
SPE	Solid phase extraction
STEM	Scanning transmission electron microscopy
t_0	Dead time
TEM	Transmission electron microscopy
TFA	Trifluoroacetic acid
TGA	Thermogravimetric analysis
THF	Tetrahydrofuran
TLC	Thin-layer chromatography
TPABr	Tetrapropyl ammonium bromide
t_R	Retention time
u	Linear velocity flow rate
UTLC	Ultrathin layer chromatography
XPS	X-ray photoelectron spectroscopy
α	Selectivity

Chapter 1
Introduction

Abstract This chapter introduces the principles and recent development of stationary phases in liquid chromatography, thin-layer chromatography and solid phase extraction, as well as substrates for laser desorption/ionization mass spectrometry.

1.1 Concepts and Definitions in Chromatography

Evaluating the performance of chromatographic separation is important for the optimization and comparison of different methods. Common performance parameters are defined and discussed here.

The retention of solutes on a stationary phase is determined by the measurement of retention time, t_R, or retention factor, k, in liquid chromatography (LC) and retardation factor, hR_F, in thin-layer chromatography (TLC).

$$k = (t_R - t_0)/t_0 \text{in LC}, \quad (1)$$

where t_0 is the dead time.

$$hR_F = 100 \times (Z_X/Z_S) \text{ in TLC}, \quad (2)$$

where Z_X is the distance traveled by the analytes and Z_S is the distance traveled by the solvent.

The difference in retention is described with selectivity, α.

$$\alpha = k_1/k_2 \text{ in LC}, \quad (3)$$

$$\text{or} \quad \alpha = hR_{F1}/hR_{F2} \text{ in TLC}, \quad (4)$$

T. Lu, *Nanomaterials for Liquid Chromatography and Laser Desorption/Ionization Mass Spectrometry*, Springer Theses,
DOI: 10.1007/978-3-319-07749-9_1,

where k_1 and k_2 or hR_{F1} and hR_{F2} describe the retention of analytes 1, and 2. k_1/hR_{F1} is larger than k_2/hR_{F2}. Therefore, the value of α is equal or larger than 1. A large α means large difference in retention on the studied stationary phases for the two analytes of interest.

The separation efficiency is characterized by theoretical plate number, N, and theoretical plate height, H.

$$N = 5.54 \times (t_r/w)^2, \tag{5}$$

where w is the peak width at half height.

$$H = L/N, \tag{6}$$

where L is the length of the column for LC or the distance traveled by the analyte in TLC. This treatment of separation efficiency assumes that the chromatographic band has a Gaussian peak shape.

1.2 Stationary Phases in Liquid Chromatography

1.2.1 Stationary Phases in Normal-Phase Thin Layer Chromatography

Thin layer chromatography (TLC) is used widely in organic synthesis, food and pharmaceutical industry, and environment analysis. This technique is simple, fast and inexpensive compared to other chromatographic methods. In addition, TLC is relatively simple to perform and does not require extensive training before one can conduct TLC separation.

Similar to other chromatographic techniques, normal-phase and reversed-phase retention mechanisms are used in TLC. While researchers using TLC spend most of their time optimizing the mobile phase conditions, the choice of the stationary phase is crucial to a successful TLC separation. Therefore, the development of various stationary phases that meet the separation requirements for different purposes is critical. Commonly-used TLC stationary phases include silica gel, alumina, cellulose and chemically bonded silica gel. Silica gel, alumina and cellulose are polar phases and used for normal-phase separation. Chemically bonded silica stationary phases are typically hydrophobic and used for reversed-phase separation. Here we focus on the normal-phase TLC separation.

Silica gel. Silica gel is the most widely used stationary phase in TLC separation. Silica gel is commercially-available and provides good separation to polar analytes. Silica gel is available in different sizes, pore structures, surface properties and surface areas. The various types of silica gel provide the possibility to meet different application purposes.

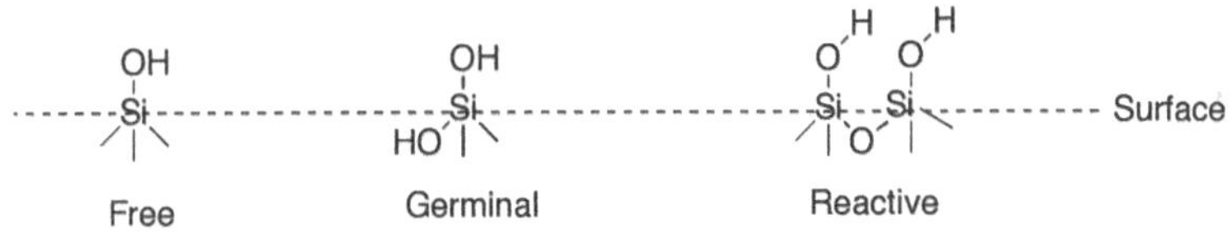

Free Germinal Reactive Surface

Fig. 1.1 Structures of surface silica

Figure 1.1 shows the structure of the silica gel surface. The surface of silica gel is fully hydroxylated, four or five SiOH, silanol, groups per square nanometer [1]. The hydroxyl groups interact through hydrogen bonding and other polar interactions with the analytes. There are three types of surface silanol groups: free, germinal and reactive (Fig. 1.1). The relative energy of interaction with these silanol groups is germinal < free < reactive [2]. The amount of the surface hydroxyl groups can be changed by dehydration treatment under high temperatures [3]. The adsorbed water can be removed at relative low temperature. The silanol groups can be converted to siloxane groups under higher temperature. Silica gel that is processed under different temperatures shows different retentions to analytes because of the presence of different types of surface silanol groups. Due to the difference in the energy of adsorption, the separation efficiency is limited. In addition, the metal impurities in silica gel can active the adjust silanol group and as a result the silanol group is more acidic [4].

Aluminum oxide. Aluminum oxide (alumina) is another type of stationary phase that is commonly-used in TLC. Similar to silica gel, aluminum oxide is a polar phase and contains hydroxyl groups on the surface. It can have hydrogen bond and polar interactions with analytes. For many analytes, alumina shows similar selectivity to silica gel. However, the alumina phase can have three forms depending on the treatment processes: acidic, neutral and basic. Therefore, it can also have electrostatic interactions with analytes if the surface hydroxyl groups are ionized.

Similar to silica gel, the amount of surface hydroxyl groups can be changed by heat treatment. The alumina heated at 400 °C loses the adsorbed water and has six hydroxyl groups per square nanometer of surface [3]. If alumina is heated above 800 °C, the surface hydroxyl groups are completely removed [5].

Another difference between alumina and silica is that alumina has Lewis acid and Lewis base centers while silica does not. The strong Lewis acid centers interact with polar and unsaturated analytes; the strong Lewis base centers interact strongly with acidic analytes [6].

A survey of common normal-phase stationary phases for TLC. Cellulose is also a common stationary phase in TLC. It is a polymer made mainly from plants. The structure of cellulose is shown in Fig. 1.2. Compared to silica and alumina, cellulose also has hydroxyl groups on the surface, which can provide polar and hydrogen bond interactions. It is best used for the separation of hydrophilic analytes using normal-phase retention mechanisms [7]. Similar to cellulose, starch and sucrose have also been used as TLC stationary phases [8, 9]. The application of sucrose as a stationary phase is limited by its high solubility in water [9].

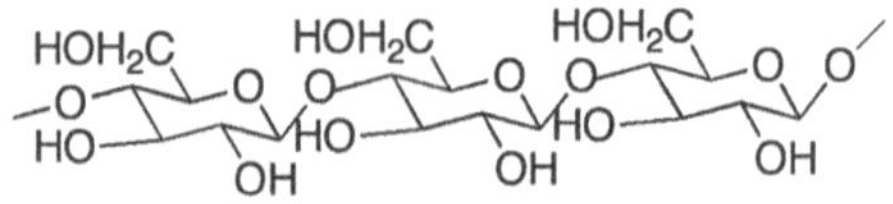

Fig. 1.2 Structure of cellulose

Magnesium oxide has been used for the separation of aromatic analytes, such as carotenes [10]. Zinc carbonate has been used to separate aldehydes and ketones due to their strong interactions with carbonyl groups [11]. Mixed stationary phases have also been studied. For example, charcoal and silica gel have been mixed for the analysis of ketones [11]. Silica and alumina have also been used as a mixed phase for the separation of sugars [12].

HPTLC. High performance TLC (HPTLC) provides higher efficiency than conventional TLC plates. Therefore, the separation is faster and provides high sensitivity. Chromatographic separation efficiency is related to the particle size and homogeneity. Van Deemter equation describes the origins of band broadening:

$$H = A + B/u + Cu, \quad (7)$$

where u is the linear velocity flow rate, the A coefficient describes the band dispersion associated with analytes moving through multiple flow paths, the B coefficient describes the band dispersion caused by longitudinal diffusion, and the C coefficient describes the resistance to mass transfer. The C coefficient is actually controlled by at least two different resistance to mass transfer interactions, including C_m and C_s, representing the resistance to mass transfer in a mobile phase and a stationary phase, respectively.

$$A = 2\lambda d_p, \quad (8)$$

where λ is a constant that is related to the heterogeneity of the stationary phase and the packing condition; d_p is the diameter of the particle size.

$$B = 2\gamma D_m, \quad (9)$$

where γ is a constant for packed column. D_m is the binary diffusion coefficient for the analyte diffusing in the mobile phase.

$$C_m \infty\ d_p^2/D_m, \quad (10)$$

$$C_s \infty\ d_f^2/[(\gamma + 1)D_s], \quad (11)$$

where d_f is the film thickness of the stationary phase and D_s is the diffusion coefficient in the stationary phase.

From the above equations, the A term and C_m term are related to the particle size, d_p. By decreasing the particle size, H can be decreased and the efficiency is improved. Efficient HPTLC separation is achieved by using particles with smaller diameters as stationary phases. Conventional TLC stationary phases use coarse-grained particles with diameter of 20–40 μm with wide size distribution, while HPTLC stationary phases have particle diameters about 5–10 μm [3]. It has been reported that using 10 μm particles can improve the separation efficiency compared to conventional TLC [13]. Due to the improved efficiency, the separation using HPTLC is more rapid and the length of the plate required is shorter. Commonly-used HPTLC stationary phases include silica gel and cellulose.

UTLC. Ultra-thin layer chromatography (UTLC) is a relatively new technique. Compared to conventional TLC and HPTLC, UTLC uses a much thinner stationary phase. The thickness of conventional TLC mats is about 100–250 μm while the thickness of UTLC stationary phases is about 5–30 μm.

The first UTLC application was reported by Hauck et al. [14]. Silica monolith with thickness of 10 μm was used as the stationary phase. The silica monolithic UTLC plate showed much faster separation compared to conventional TLC. Because the thickness of the mat is thin, the amount of mobile phase needed is much less than conventional TLC. It is a great advantage if large amount of TLC analysis needs to be performed. The silica monolith did not show significant improvement in efficiency [14].

The Olesik group reported multiple UTLC papers using electrospun nanofibers as stationary phases since 2009 [15–18]. From the above discussion about efficiency using van Deemter equation, by using nano-scaled materials improvements in the separation efficiency are expected.

The preparation of electrospun UTLC plate is simple, fast and low cost. Electrospinning usually takes from 5 min to 2 h. The apparatus of electrospinning is simple and cost effective. When a high electric field is placed between the polymer drop at the end of the syringe tip and the grounded collector, the surface of the polymer solution becomes charged by the electrical field induced migration of free charges through the liquid [19]. When the potential difference is large enough to overcome the surface tension of the drop, a Taylor cone is formed. Following the creation of the Taylor cone, fibers are ejected toward the conductive collector in the form of a polymer jet [20]. As the polymer jet moves toward the collector in a chaotic pattern, the fibers are pulled by the force of the applied field. The net result is nanofibers of a controlled diameter, typically in the 100–1,000 nm range, as determined by the potential difference, solution flow rate and viscosity, polymer molecular weight, and temperature, as well as, chamber humidity.

The electrospun UTLC stationary phases show many advantages compared to conventional TLC plates. The stationary phase is composed of electrospun nanofibers adhering on aluminum or stainless steel plates. No binder is required. The stationary phase is homogenous. Polyacrylonitrile [15], glassy carbon [16] and PVA [18] electrospun UTLC plates have been reported. These materials with different functionalities provide various selectivities. Moreover, the nanostructured stationary phases provide significantly improved efficiency. Aligned electrospun UTLC plate has also been prepared [17]. The advantage of aligned fibers is that the analysis speed is greatly improved. Additionally, similar to monolithic UTLC, electrospun UTLC also provides sensitive analysis and improved detection limit. Electrospun UTLC with photoluminescence indicator has also been reported recently [21].

Other UTLC stationary phases have also been reported. For example, a polymer monolithic stationary phase has been used for the separation of peptides and can be used for on-plate matrix-assisted laser desorption/ionization (MALDI) detection [22]. UTLC-MALDI-MS analysis using polyacrylamide coated silica as a stationary phase was also used for the characterization of peptides [23]. Nanostructured mats prepared using the glancing angle deposition (GLAD) technique has been reported [24–28]. Silver nanorods array has been reported as a stationary phase for UTLC and surface enhanced Raman spectroscopy has been used for the detection [29].

1.2.2 Carbon Stationary Phases in HPLC

Silica based stationary phases are the most commonly-used in HPLC. The silica gel particles are commercially-available in different sizes and porosities. However, silica gel is not stable under extreme pH conditions. For example, silica gel dissolves at pH < 2. At pH > 8, the siloxane bond that connects the chemical functionalities to the bare silica dissociates.

Carbon stationary phase has been widely used in HPLC. Compared to silica-based phases, carbon provides unique selectivity and high stability under extreme pH conditions and high temperatures. However, the use of a carbon phase was not successful until Knox et al. [30] introduced the preparation method of a porous graphitic carbon phase. After that Carr group [31] and Olesik group [32] have reported the use of carbon coated inorganic particles, respectively, in order to improve the mechanical strength, which provides carbon particles with different surface areas and lowers the preparation cost.

Porous graphitic carbon. Porous graphitic carbon, also known as Hypercarb, is the most widely used carbon phase in HPLC. The preparation of Hypercarb involves the deposition of organic monomers on the silica template, polymerization, carbonization, removal of the silica template and graphitization [30]. The cost of the preparation of Hypercarb is high.

Carbon-clad inorganic particles. The Carr group has reported carbon-clad zirconia [31], alumina [33] and silica particles [34]. They used chemical vapor deposition (CVD) to coat the carbon on the surface of the template. This technique is less expensive than the preparation of Hypercarb. In order to grow carbon using the CVD method the availability of active sites on the template surface is required. Because silica does not have active sites, in order to use silica as a template silver needs to be coated on the silica template first [34]. The templates are not removed using the CVD method.

Low temperature glassy carbon. The Olesik group has reported low temperature glassy carbon as a stationary phase for HPLC [32]. The preparation of low temperature glassy carbon is simple and low cost. A carbon precursor, organic oligomers, was dissolved in the organic solvent before use. Then the silica or zirconia particles were dispersed in the solution containing the carbon precursor. The

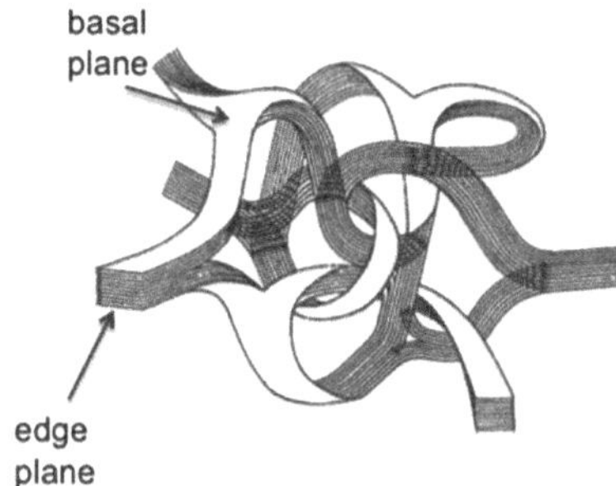

Fig. 1.3 Edge-plane and basal-plane sites in amorphous carbon [35]

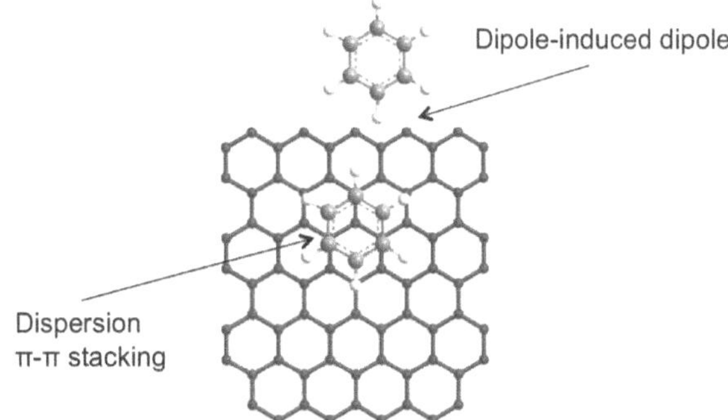

Fig. 1.4 Possible interactions between benzene and amorphous carbon

organic molecules adsorbed onto the surface of the template by weak interactions in solution. After the solvent was removed the particles were carbonized. Similar to the CVD method, the template particles were not removed for low temperature glassy carbon. The reason that it is called low temperature glassy carbon is that the carbonization temperature is low compared to that used to graphitize Hypercarb, below 700 °C versus above 2,000 °C.

Homogenous ordered carbon. The currently-used carbon phases are amorphous carbon. Amorphous carbon is composed of intertwining graphite ribbons. The alignment of the graphite ribbons is not homogenous. A representative structure of amorphous carbon is shown in Fig. 1.3 [35].

Amorphous carbon has at minimum two sites of interactions: basal-plane sites and edge-plane sites. The basal-plane surface is where a graphite layer is parallel to the surface. Edge-plane sites are where the edges of graphite layers are exposed to the surface. The two different sites have different polarity. Edge-plane sites are more polar than basal-plane sites. Additionally, edge-plane sites may contain different functionalities [36]. Due to the difference between the two types of sites, different interactions are expected. For example, if benzene is used as an analyte on a carbon phase, there are at least two types of interactions (Fig. 1.4). If benzene interacts with an edge-plane site the interaction may be dipole-induced dipole interaction. If benzene interacts with a basal-plane site it is more likely that the interaction is dispersive interaction/π-π interaction. Unfortunately, no homogenous

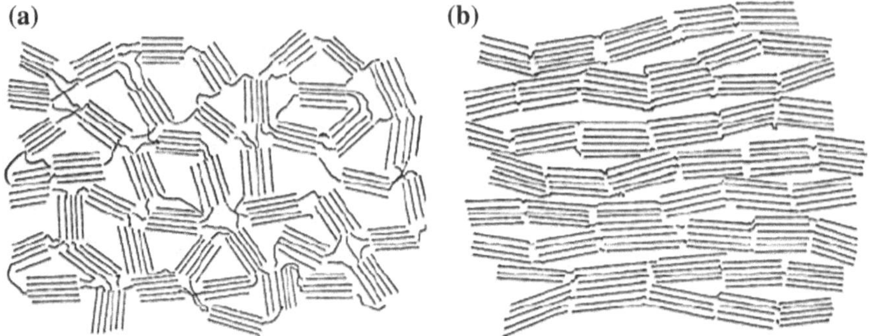

Fig. 1.5 Representative models of **a** amorphous carbon and **b** graphitic carbon [38]

ordered carbon containing only basal-plane or edge-plane carbon for chromatography application has been reported yet.

Hypercarb is used as a carbon phase with a more homogenous surface. The last step of the preparation of Hypercarb is graphitization under high temperature. After graphitization the alignment of the graphite layers becomes more ordered (Fig. 1.5) [37] . Therefore, Hypercarb is regarded to be composed with a large amount of basal-plane sites and a small amount of edge-plane sites. Due to the high temperature treatment under reduced atmosphere most of the organic functionalities on edge-plane sites are removed.

1.3 Sorbents in Solid Phase Extraction

Solid phase extraction (SPE) is a simple technique of separation methods. It uses solid media to adsorb the analytes of interest in solution. Firstly the analyte solution containing the interference is loaded to the SPE sorbent. Then the interference is eluted with a solvent and the analyte stays adsorbed on the sorbent. Finally, the analyte is eluted with a strong solvent.

The sorbents used for SPE are very similar to the stationary phases of liquid chromatography. SPE sorbents include normal-phase and reversed-phase sorbents. Chemically-modified silica-based sorbents are the mostly widely used sorbents in SPE. Different functionalities can be attached to the surface of silica, such as octyl, octadecyl and cyano groups. The drawback of chemically-modified silica-based sorbents is that the remaining free silanol groups after the chemical modification can interact with the sample solution. These silanol groups are polar and acidic, which causes the complexity of the extraction process. The end-cap method has been used to solve this problem. This method uses bulky functional groups to react with the free silanols. Although there are free silanols remaining on the surface the bulky functional group can block the free silanols sterically. As a result, they cannot interact with the samples.

Carbon sorbents have also been used in SPE. Carbon can interact with aromatic analytes more strongly through π-π interactions. In addition, carbon is more stable

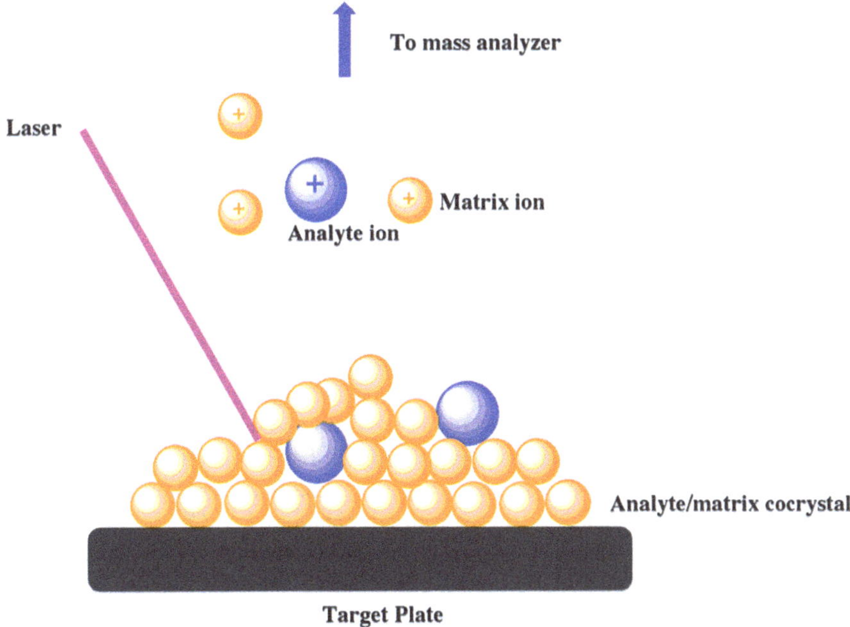

Fig. 1.6 A diagram of the mechanism of MALDI

and chemically inert. As a result of the development of carbon nano materials, such as carbon nanotubes, highly efficient extraction using carbon nano materials has been achieved [39].

1.4 Substrates for Laser Desorption/Ionization Mass Spectrometry

MALDI. MALDI is the most widely-used technique in laser desorption/ionization mass spectrometry. It is a simple and fast method to ionize and analyze molecules with large molecular weights and thermal labile analytes [40, 41]. MALDI uses an organic matrix to absorb the laser energy and transfer the energy to the analytes (Fig. 1.6). The organic matrix can be mixed with the analyte solution or it can be pre-coated on to the target plate [42, 43]. After the sample solution is spotted on to the plate the analyte and the matrix form cocrystals. The formation of cocrystals is required for MALDI analysis. Homogeneous sample distribution is critical for the following step. If the sample distribution is homogeneous the searching of "sweet spot" is less time consuming [42]. More importantly, a homogenous sample distribution is required for quantitative analysis [44]. The selection of the type of matrix is also important. However, there is no theoretical prediction method for the selection of matrix. Currently, trial and error method is still used [45]. Another limitation of MALDI is that the organic matrices generate signals as well. These signals cause low mass interferences in the spectra [45].

SALDI. Surface-assisted laser desorption/ionization (SALDI) has been developed as an alternative to MALDI MS. SALDI uses a nanostructured inorganic material in place of an organic matrix in MALDI [45]. The SALDI materials usually have intensive UV absorption, such as carbon [45], silicon [46] and metal nanoparticles [47]. The SALDI material can absorb the laser energy and transfer the energy to the analytes, similar to organic matrices. The nanostructure is critical. For the same material, such as TiN, the bulk form cannot generate any ion signals while nanostructured TiN can be used for SALDI successfully [48]. The exact reason is not clear yet. However, it may relate to two reasons. (1) The nanostructured surface has much higher surface area and the area under the laser spot becomes much larger compared to the bulk form [49]. (2) The nanostructured surface can hold the analyte molecules in pores. It has been found that the signal intensity is highly related to the pore structure of the SALDI materials [50].

PALDI. Polymer-assisted laser desorption/ionization (PALDI) has been reported by Roeraade, et al. [51, 52] Aromatic polymers and oligomers have been used as PALDI matrixes. The PALDI matrix can absorb laser energy and transfer it to the analytes. Due to the high molecular weight of PALDI matrixes, they are not ionized when the laser with low energy for small molecules is applied. Therefore, the spectra are clean and there is no low mass interference. However, the PALDI application is limited to the analytes with low molecular weight.

SELDI. Surface-enhanced laser desorption/ionization (SELDI) is a technique that is good for complex samples. The use of organic matrix is required in SELDI. A polymeric substrate is pre-coated on to the target plate. The substrate is usually hydrophobic and interacts with protein strongly. SELDI provides an additional step of on-plate purification of the sample. If the analyte of interest interacts with the SELDI substrate more strongly than other impurities, such as salts, the impurities can be washed away from the sample spot while the analyte of interest stays remained on the target plate. The spectra become much cleaner and simple to interpret [53, 54].

ME-SALDI. Matrix-enhanced SALDI has been reported by He et al. [55, 56] This method combines MALDI and SALDI. Both organic matrix and SALDI material are used. The advantage is that the matrix signals are suppressed, the degree of fragmentation is lowered and the spectrum is cleaner in the low mass region. The SALDI material is responsible for the energy absorption and transfer. The organic matrix desorbs and carries the analytes into the gas phase and the amount of analytes in the gas phase is increased. The organic matrix also absorbs the extra energy, which prevents the analytes from fragmentation. In addition, the organic matrix also serves as a proton source that assists the ionization of the analytes. ME-SALDI has been successfully applied to metabolite imaging and small molecule analysis [57].

References

1. Scott RP, Cucera P (1975) J Chromatogr Sci 13:337
2. Snyder LR, Ward JW (1966) J Phys Chem 70:3941
3. Grinberg N (1990) Modern thin-layer chromatography. Marcel Dekker Inc, New York
4. Nowrocki J (1997) J Chromatogr A 656:353
5. Peri JB (1965) J Phys Chem 69:211
6. Snyder LR (1964) J Chromatogr 16:55
7. Fried B, Sherma J (1994) Thin-layer chromatography: techniques and applications, 3rd edn. Marcel Dekker Inc, New York
8. Davidek J (1964) In: Marini-Bettolo GB (ed) Thin layer chromatography. Elsevier, New York, pp 340
9. Woolenweber P (1969) In: Stahl E (ed) Thin layer chromatography. A laboratory handbook, 2nd edn. Springer, New York, pp 32
10. Strain HH, Svec MA (1969) Adv Chromatogr 8:119
11. Rossler H (1969) In: Stahl E (ed) Thin layer chromatography. A laboratory handbook, 2nd edn. Springer, New York, pp 23
12. Wiedenhof N (1964) J Chromatogr 15:100
13. Gocan S, Ursu E (1980) J Chromatogr 198:421
14. Hauck HE, Bund O, Fischer W, Schulz MJ (2001) Planar Chromatogr 14:234
15. Clark J, Olesik SV (2009) Anal Chem 81:4121
16. Clark J, Olesik SV (2010) J Chromatogr A 1217:4655
17. Beilke MC, Zewe JW, Clark JE, Olesik SV (2013) Anal Chim Acta 761:201
18. Lu T, Olesik SV (2013) J Chromatogr B 912:98
19. Reneker DH, Yarin AL, Fong H (2000) J Appl Phys 87:4531
20. Ramakrishna S, Fujihara K, Teo W, Lim T-C, Ma Z (2005) An introduction to electrospinning and nanofibers. World Scientific Publishing Co. Pte.: Toh Tuck Link, Singapore
21. Kampalanonwat P, Supaphol P, Morlock GE (2013) J Chromatogr A 1299:110
22. Salo PK, Vilmunen S, Salomies H, Ketola RA, Kostiainen R (2007) Anal Chem 79:2101
23. Zhang Z, Ratnayaka SN, Wirth MJ (2011) J Chromatogr A 1218:7196
24. Jim SR, Taschuk MT, Morlock GE, Besuidenhout LW, Schwack W, Brett M (2010) J Anal Chem 82:5349
25. Oko AJ, Jim SR, Taschuk MT, Brett MJJ (2011) Chromatgr A 1218:2661
26. Jim SR, Oko AJ, Taschuk MT, Brett MJJ (2011) Chromatgr A 1218:7203
27. Hall JZ, Taschuk MT, Brett MJJ (2012) Chromatgr A 1266:168
28. Jim SR, Foroughi-Abari A, Krause KM, Li P, Kupsta M, Taschuk MT, Cadien KC, Brett MJJ (2013) Chromatgr A 1299:118
29. Chen J, Abell J, Huang Y, Zhao Y (2012) Lab Chip 12:3096
30. Knox JH, Kaur B, Millward GR (1986) J Chromatogr 352:3
31. Weber TP, Carr PW (1990) Anal Chem 62:2620
32. Engel TM, Olesik SV, Callstrom MR, Diener M (1993) Anal Chem 65:3691
33. Paek C, McCormick AV, Carr PW (2010) J Chromatogr A 1217:6475
34. Paek C, McCormick AV, Carr PW (2011) J Chromatogr A 1218:1359
35. Jenkins GM, Kawamura K (1971) Nature 231:175
36. Pereira LJ (2008) Liq Chromatogr RT 31:1687
37. Franklin RE (1951) Proc Roy Soc A209:196
38. Harris PJF (2005) Crit Rev Solid State Mater Sci 30:235
39. Herrera AV, Gonzalez-Curbelo MA, Hernandes-Borges J, Rodriguez-Delgado MA (2012) Anal Chim Acta 734:1
40. Tanaka K, Waki H, Ido Y, Akita S, Yoshida Y, Yohida T (1988) Rapid Commun Mass Spectrom 2:151
41. Karas M, Bachmann D, Hillenkamp F (1985) Anal Chem 57:2935
42. Vorm O, Roepstorff P, Mann M (1994) Anal Chem 66:3281

43. Axelsson J, Hoberg A-M, Waterson C, Myatt P, Shield GL, Varney J, Haddleton DM, Derrick PJ (1997) Rapid Commun Mass Spectrom 11:209
44. Jaskolla TW, Karas M, Roth U, Steinert K, Menzel C, Reihs K (2009) J Am Soc Mass Spectrom 20:1104
45. Sunner J, Dratz E, Chen Y-C (1995) Anal Chem 67:4335
46. Wei J, Buriak JM, Siuzdak G (1999) Nature 399:243
47. Chen WT, Chiang CK, Lee CH, Chang HT (2012) Anal Chem 84(4):1924
48. Schuerenberg M, Dreisewerd K, Hillenkamp F (1999) Anal Chem 71:221
49. Buryak AK, Serdyuk TM (2011) Prot Met Phys Chem Surf 47:911
50. Jang J, Bae J (2005) Chem Commun 1200
51. Woldegiorgis A, Kieseritzky F, Dahlstedi E, Hellberg J, Brinck T, Roeraade J (2004) Rapid Commun Mass Spectrom 18:841
52. Woldegiorgis A, Lowenhielm P, Bjork A, Roeraade J (2004) Rapid Commun Mass Spectrom 18:2904
53. McComb ME, Oleschuk RD, Manley DM, Donald L, Chow A, O'Neil JDJ, Ens W, Standing KG, Pereeault H (1997) Rapid Commun Mass Spectrom 11:1716
54. Zaluzec EJ, Gage DA, Allison J, Watson JT (1994) J Am Soc Mass Spectrom 5:230
55. Liu Q, Xiao Y, Pagan-Miranda C, Chiu YM, He L (2009) J Am Soc Mass Spectrom 20:80
56. Liu Q, He L (2009) J Am Soc Mass Spectrom 20:2229
57. Liu Q, Xiao Y, He L (2010) Mass spectrometry imaging: principles and protocols. In: Rubakhin SS, Sweedler JV (eds) Methods in molecular biology. Humana Press, Clifton, pp 243

Chapter 2
Electrospun Polyvinyl Alcohol Ultra-Thin Layer Chromatography of Amino Acids

Abstract Electrospun crosslinked polyvinyl alcohol (PVA) was prepared and used as stationary phase for ultrathin layer chromatographic (UTLC). The nanostructured stationary phase shows greatly enhanced efficiency compared to conventional HPTLC plate. An example of using PVA-UTLC plate for the analysis of hydrolysis products of aspartame in diet coke, aspartic acid and phenylalanine, is included.

2.1 Introduction

Compared to HPLC, the range of stationary phase types is smaller for TLC. Many stationary phases can be readily developed by using electrospinning technique. In this study, polyvinyl alcohol (PVA) was used as the stationary phase for UTLC. PVA is a nontoxic, biodegradable, and a biocompatible polymer [1]. The hydroxyl groups render PVA as polar stationary phase. Because PVA nanofibers are soluble in water this would limit its use for the separation of polar compounds. Therefore efforts to enhancing the water resistance of the PVA were necessary. The most popular way of decreasing PVA's solubility in water is to crosslink the polymer. Crosslinking using high energy ionization radiation [2], and UV irradiation of PVA [3], has been used but these methods require special instrumentation or modification of the PVA polymer backbone. Crosslinking reactants such as glutaraldehyde [4, 5], glyoxal [6], formaldehyde [7], and maleic anhydride or maleic acid [7, 8], were used previously. Crosslinking of PVA nanofibers has also been accomplished after electrospinning by immersion into a solution of crosslinker [9, 10]. Another means of stabilizing PVA is to add the crosslinker to the electrospinning solution which allows electrospinning and crosslinking to occur in the same step [11].

The separation of amino acids is important in agriculture, food analysis and biomedical applications. HPLC, CE, and GC are used for the amino acids analysis [12]. However, TLC has the added benefits of simplicity, speed of analysis, and

T. Lu, *Nanomaterials for Liquid Chromatography and Laser Desorption/Ionization Mass Spectrometry*, Springer Theses,
DOI: 10.1007/978-3-319-07749-9_2,

cost-effectiveness compared to the instruments mentioned above. Therefore, TLC is also a useful and commonly-used technique for the amino acids analysis [13–16]. For example, Liu et al. [17] reported chiral separation of racemic amino acids on a TLC plate made by PVA film containing DNA. Simon et al. [18] recently compared the differences in the distribution of free amino acids in sanguine plasma for normal patients and those with different stages of diabetes, renal syndrome and hepatic cirrhosis using TLC analysis. A biodegradable UTLC device for such biomedical analyses would likely be advantageous for the diagnoses and monitoring of disease states. A small UTLC device could be more efficient and less organic solvent is needed.

The separation of amino acids using TLC has been challenging. The available stationary phases are limited as mentioned therefore complex mobile phase mixtures including a variety of organic solvents [19] are needed and additives such as acids and bases [20], metal ions [21], or surfactants [22] are usually required to improve the separation. The complex multicomponent mobile phases require time consuming optimization and some of the reagents are not usually available in regular labs. Glassy carbon UTLC has been used for amino acid separation with high efficiency [5]. However, the limitation is amino acids cannot be analyzed directly. Labeling with fluorescent dyes was required using glassy carbon UTLC. And the detection of labeled amino acids required an additional step: solvent extraction of the labeled amino acids. Otherwise the fluorescence is completely quenched by carbon. In this work we used electrospun PVA as a stationary phase which showed high efficiency and unique selectivity. With highly efficient separation, compounds can be separated even at lower selectivity which simplifies the optimization of selectivity using different mobile phases. In this work, the separation of amino acids using ternary mixtures of commonly-used organic solvents, butanol, methanol or ethyl acetate, and water, without any other additives was achieved.

Because amino acids cannot be visualized by exposure to UV light, a fluorescent label was attached to the amino acids before the separation [15]. Dye-labeled amino acids are colored and show fluorescence under UV light. The spots can be visualized during the separation and fluorescence can be detected for small quantities of the analytes. Therefore it is convenient to study the chromatographic behavior using the fluorescent labeled amino acids. Isothiocyanate derivatizing agents are commonly-used for visualization of amino acids by reacting with the amino acid to form fluorescent thioureas. Also, ninhydrin as a visualization reagent has been widely used for amino acids detection on TLC plate. We studied and separated fluorescein isothiocyanate (FITC) derivitized amino acids [23] and visualized amino acids with ninhydrin to detect amino acids on an electrospun PVA plate. The hydrolysis products of aspartame in diet coke, aspartic acid and phenylalanine, were also analyzed using PVA-UTLC plate.

2.2 Experimental

2.2.1 Materials

The polymer solution was prepared by dissolving PVA (99 + % hydrolyzed, M_w 89,000–98,000, Sigma, St. Louis, MO) in distilled water maintained at 80 °C. Glutaraldehyde (GA, 25 % in water, Baker) and hydrochloric acid (0.1 M) were used for crosslinking the PVA. Alanine (ala), methionine (met), aspartic acid (asp), glutamine (gln), arginine (arg) mono hydrochloride acid, tyrosine (tyr), phenylalanine (phe), histidine (his), and threonine (thr) were obtained from Sigma (St. Louis, MO). Cysteine (cys) and tryptophan (trp) were purchased from Aldrich (Milwaukie, WI). Merck silica HPTLC plates on glass plate with 4–8 μm diameter silica particles and 150–200 μm film thickness were obtained from EMD Chemicals (Gibbstown, NJ).

2.2.2 Electrospinning

Electrospinning of the polymer was accomplished using a previous reported procedure [11] with modifications and the instrumental setup is shown in Fig. 2.1. A syringe pump (KD Scientific, model: 780100), high voltage power supply (Spellman, Hauppauge, NY, model: CZE1000PN30), a stainless steel collector (11 cm × 12 cm) covered with aluminum foil (super strength, Reynolds), and a Plexiglas enclosure. An 8 % (w/w) aqueous PVA solution was optimized to be the concentration for electrospinning. Before the electrospinning, glutaraldehyde (GA:PVA, mol:mol, 90:1) and HCl (HCl:GA, mol:mol, 1:5) were added to start the crosslinking reaction [12]. The feed rate of the PVA polymer solution was 0.5 mL/h. The voltage applied was 20 kV. The distance from the spinneret needle tip to the collector was optimized to be 20 cm. The relative humidity was controlled to 30 % or below by purging the plastic enclosure with dry N_2 while electrospinning and was monitored by VWR Traceable® digital hydrometer. At relative humidity levels higher than 30 %, bead formation on the top of the fibers was noted. The time of electrospinning for each polymer solution was 30 min.

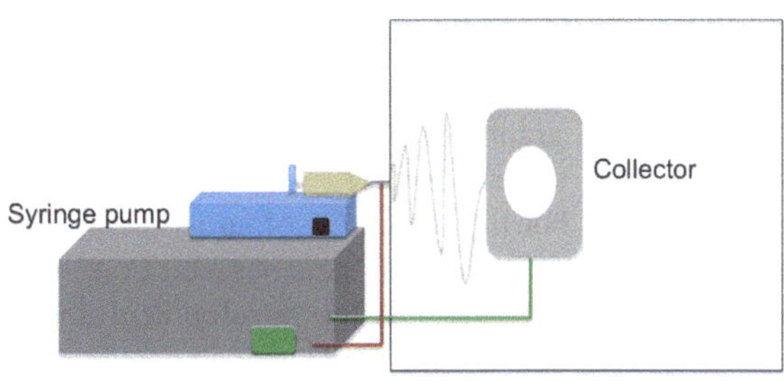

Fig. 2.1 Instrument setup used for electrospinning

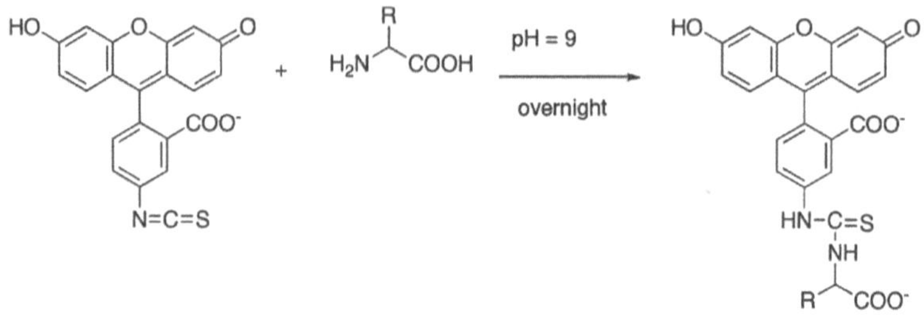

Fig. 2.2 Reaction scheme of the synthesis of FITC-labeled amino acids

2.2.3 FITC Labeled Amino Acids

The amino acids were labeled with FITC dye using the published reaction conditions [23, 24]. The reaction is shown in Fig. 2.2. 0.5 mL amino acid solution (2 mg/mL each, in 0.2 M pH = 9 carbonate-bicarbonate buffer) were mixed with 0.05 mL (60 μmol/mL in acetone with trace amount of pyridine) FITC (Sigma, St. Louis, MO). The reaction solution was protected from light for overnight and was stored in the refrigerator. The FITC-labeled amino acids were used directly and were stable for analysis for 2 weeks. The concentration of the FITC-labeled amino acids was calculated from the amount of FITC added which was the limiting reagent in this reaction.

2.2.4 Thin Layer Chromatography

The electrospun PVA plate was cut into 7 cm × 3 cm UTLC devices. The FITC labeled amino acid solutions were spotted onto the bottom of UTLC plate using fused silica capillary tubes (id: 50–250 μm, Polymicro Technologies, Phoenix, AZ). The volume of the solution that was spotted onto the plate was calculated from the volume difference in the capillary tube before and after spotting. About 1 mL of the freshly prepared mobile phase was used for each development. The development chamber was an 80 mL TLC chamber. Before each development, the mobile phase and plate were equilibrated for about 10 min. The method of the analysis of the results was reported previously [25]. Briefly, the resulting spots were visualized and analyzed by a digital documentation system from Spectroline (model CC-80). The images of FITC-labeled amino acids were recorded by taking a digital photograph which was then converted to chromatograms using a TLC analyzer. The separation using a Si-Gel HPTLC plate was conducted using the same method. For the separation of amino acids visualized by ninhydrin reaction, the amino acids (5 mg/mL aqueous solution) were spotted on the bottom of electrospun PVA plate. After development the ninhydrin solution (0.3 g in 100 mL of *n*-butanol with 3 mL of acetic acid) was sprayed evenly onto the plate using a TLC reagent nebulizer (Kimble-Chase Vineland, NJ). After the plate was dry it was heated for 10 min at 110 °C [26]. Diet coke was heated and refluxed for 24 h. Then it was concentrated from 100 mL to ~1 mL. The analysis of treated diet coke was performed using the same method that has been used for unlabeled amino acid. Because the spots show different colors after ninhydrin reaction, a camera is not required to record the image. The image of the separation using ninhydrin as a visualization reagent was scanned using an EPSON GT-2500 scanner and the chromatogram was obtained using ImageJ 1.43u software. The chromatographic parameters were analyzed using PeakFit^{TM} (version 4, SPSS Inc.). All of the results were based on at least three measurements.

2.2.5 Instrumentation

The micrographic images were obtained with a Hitachi S-4300 (Hitachi High Technologies, America, Inc., Pleasanton, CA) scanning electron microscope (SEM). Before the SEM analysis the sample was sputter coated with gold for 2 min at 10 μA. Vertical SEM sample stage was used for mat thickness measurement. The thickness was measured from the image of the cross-section of the plate.

2.3 Results and Discussion

2.3.1 Electrospinning for Optimization of PVA Nanofibrous Mat Structure

Two methods of producing water resistant PVA nanofibers were tested: cross-linking during electrospinning and cross-linking after electrospinning. Tang et al. [11] studied in situ crosslinking of PVA while electrospinning previously. An optimal amount of the GA, crosslinker, and the HCl, catalyst, was determined in previously for the best water resistance and minimized swelling when the PVA fibers were exposed to water. Here we optimized the electrospinning concentration, 8 % PVA in water, to generate beads free fibers. The water resistance of in situ crosslinked PVA using these optimized conditions was compared to that using extensive crosslinking of PVA after electrospinning. For the post-electrospinning reaction, the PVA fibrous mat is immersed into a solution prepared by dissolving 75 % GA in acetone (1 %, v:v). The crosslinking reaction was stopped after 1, 3, and 5 h, respectively. Using the immersion method, the fibers partially dissolved in water which was the solvent of the crosslinker GA. On the other hand, the in situ crosslinked PVA nanofibers maintained the morphology quite well after crosslinking. Figure 2.3 shows the SEM image of the morphology of the in situ

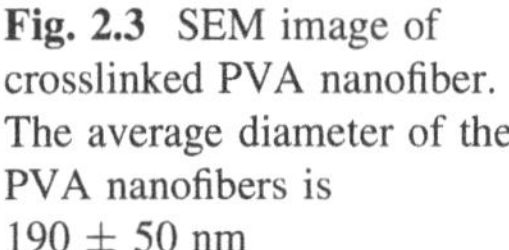
Fig. 2.3 SEM image of crosslinked PVA nanofiber. The average diameter of the PVA nanofibers is 190 ± 50 nm

crosslinked electrospun nanofibers on the aluminum foil. The average diameter of the nanofibers is ca. 200 nm. The PVA nanofibrous stationary phase adheres strongly to the aluminum foil and it was easy to cut into desired sizes.

Figures 2.4, 2.5, 2.6 illustrates the enhanced-water stability gained through in situ crosslinking electrospinning. Figure 2.4 illustrates that noncrosslinked PVA nanofibers dissolve readily when soaked for 15 min in water; Fig. 2.5 shows markedly less dissolution for the cross-linked PVA when soaked in water for 15 min and Fig. 2.6 shows that minimal change in the morphology of the cross-linked fibers occurs when soaked in the optimized mobile phase mixture for the separation of the FITC-labeled amino acids for 15 min. Clearly the in situ crosslinking was needed and effective.

When electrospinning each polymeric solution for 30 min, a nanofibrous mat with a thickness of approximately 5 µm was produced. Due to the crosslinking reaction, after 30 min the polymeric solution became too viscous to electrospin. However, one layer, 5 µm thick, electrospun PVA phase was not thick enough and even the minimum amounts of analyte solution that can be applied would overload the plate. By electrospinning new PVA solutions for 30 min each the possibility of increasing the PVA layer thickness was provided. Figure 2.7 illustrates mat thickness increases when more layers of PVA were electrospun. Similar fibrous morphology was observed for one, two, and three layers (30 min each of electrospinning) mats. Compacting of the nanofibers beneath the upper layers was not observed for the thicknesses studied. However, minor beads were observed on the four-layered mat which is consistent with previous results [25].

Figure 2.8 illustrates the variation of solvent migration distance with the square root of time. The observed linear dependence ($r^2 > 0.999$) was anticipated by the Lucas–Washburn's theory of capillary flow through porous media [27]. This dependence is similar to that observed with the other electrospun UTLC plates [25, 28].

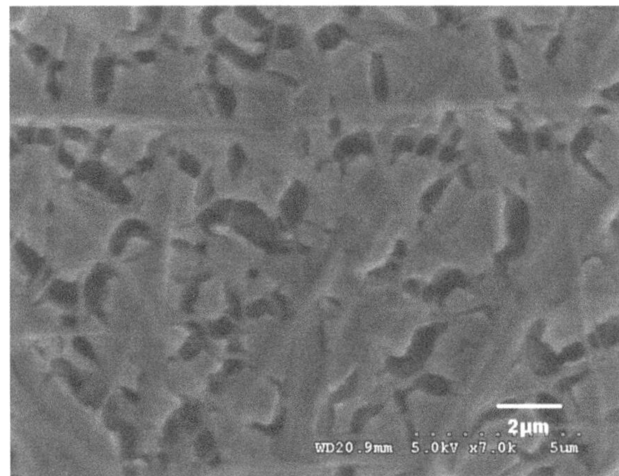

Fig. 2.4 SEM image of PVA nanofibers without crosslinking after soaked in water

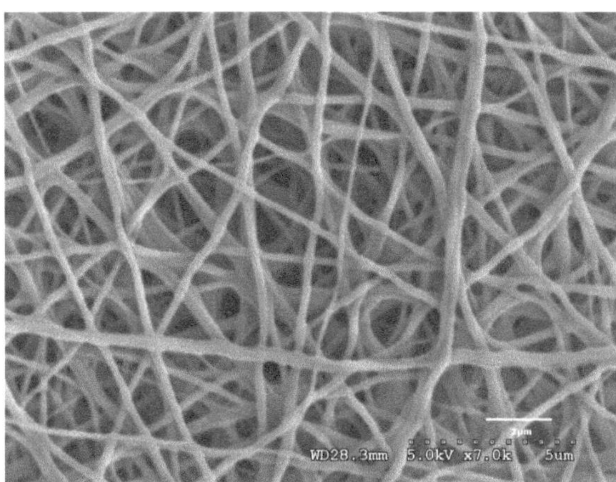

Fig. 2.5 SEM image of crosslinked PVA nanofibers after soaked in water

Fig. 2.6 SEM image of crosslinked PVA nanofibers after soaked in methanol, butanol and water (7:5:1, v/v/v)

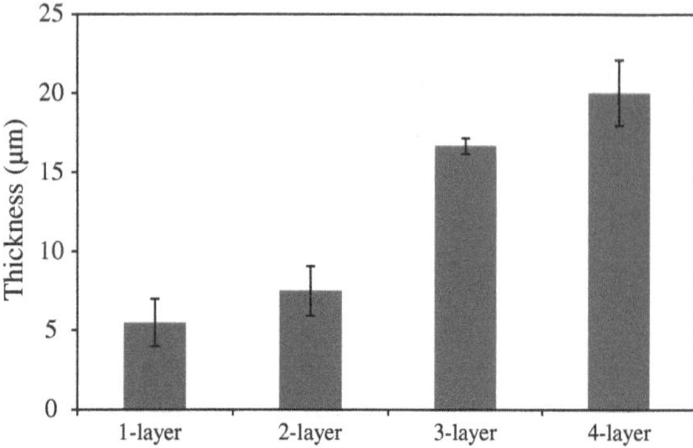

Fig. 2.7 Different mat thickness when multiple layers of electrospun PVA

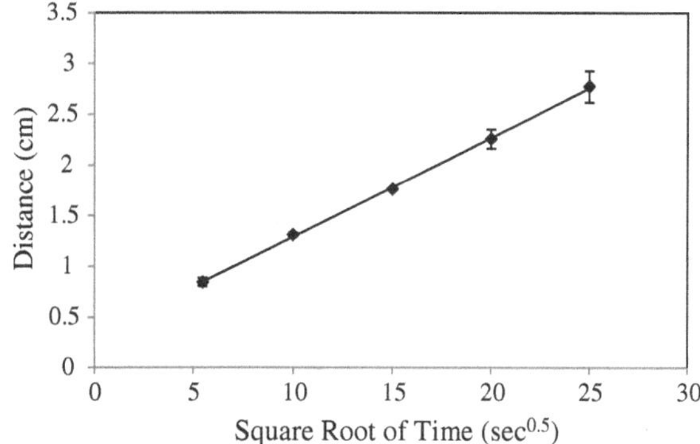

Fig. 2.8 Mobile phase (methanol/butanol/water, 7:5:1, v:v:v) migration rate on PVA plate

2.3.2 Selectivity

To optimize the mobile phase for the separation of amino acids, separations in methanol, ethanol, *n*-propanol, *i*-propanol, *n*-butanol, acetone, acetonitrile were evaluated. These mobile phases were not polar enough and all of the amino acids did not migrate up the plate. Water was added to the above solvents for further characterization. Different proportions of water were added but the selectivity among the amino acids was poor using the binary mobile phase system. Therefore, ternary mobile phase systems, including acetone/acetonitrile/water, methanol/butanol/water, ethanol/butanol/water, *n*-propanol/butanol/water, *i*-propanol/butanol/water, were tested to optimize the selectivity of the FITC labeled amino acids. Different ratios of each three components of the mobile phase were tested for all of the studied amino acids. The selectivity of the amino acids for these ternary mixtures was compared. Methanol/butanol/water (7:5:1, v:v:v) and ethanol/butanol/water (5:5:2.5, v:v:v) provided the better electivity for most of the FITC-

Table 2.1 hR_F of FITC labeled amino acids on PVA and Si-gel plate

PVA plate		Si-gel HPTLC plate
	hR_F	hR_F
Phe	34 ± 2	80 ± 2
His	24 ± 2	37 ± 7
Asp	33 ± 3	42 ± 7
Cys	11 ± 1	55 ± 1
Met	40 ± 2	77 ± 2
Thr	21 ± 2	69 ± 1
Trp	38 ± 3	27 ± 1
Gln	35 ± 3	62 ± 2
Ala	39 ± 2	73 ± 1
Tyr	28 ± 1	77 ± 2
Arg	37 ± 2	56 ± 1

labeled amino acids than the other mobile phase systems. For the solvent system of methanol/butanol/water to travel about 3 cm on the PVA plate, 10 min was required while ethanol/butanol/water mixtures took 15 min to travel the same distance. Therefore, by comparing the speed of the mobile phase moving up on the PVA plate, methanol/butanol/water was chosen to shorten the analysis time, and this solvent mixture was used for all the other studies. The mobile phase for the Si-Gel plate was optimized similarly and ethanol/butanol/water (7:5:0.5) was selected as the preferred mobile phase. Six to seven minutes were required for the mobile phase to travel about 3 cm on the Si-Gel plate. The hR_F values using mobile phase methanol/butanol/water (7:5:1, v:v:v) are listed in Table 2.1. Also, Table 2.1 shows the hR_F values for the optimized mobile phases for the Si-Gel plate.

The selectivity on a PVA UTLC plate was compared with the selectivity on Si-Gel with the respective optimal mobile phase (Figs. 2.9 and 2.10). Although both PVA and silica gel have hydroxyl groups which make them highly polar stationary phases, the retardation factors on the silica gel plate are slightly higher

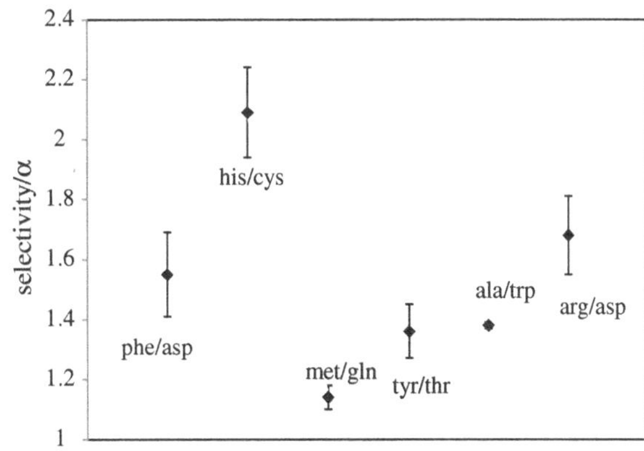

Fig. 2.9 Selectivity of FITC labeled amino acids on PVA plate (mobile phase methanol/butanol/water, 7:5:1, v:v:v)

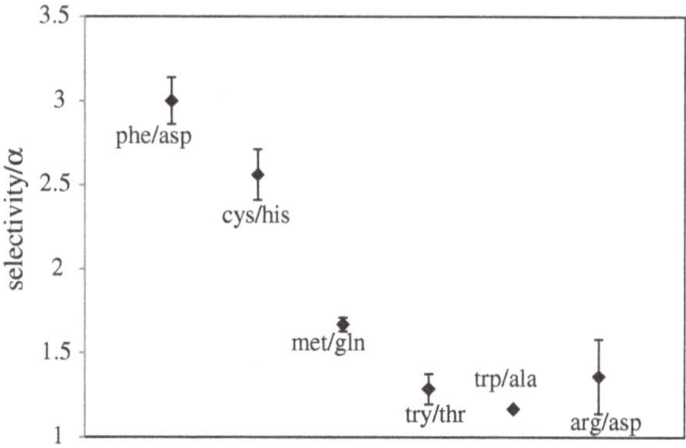

Fig. 2.10 Selectivity of FITC labeled amino acids on Si-gel HPTLC plate (mobile phase ethanol/butanol/water, 7:5:0.5, v:v:v)

than that on the PVA plate using the optimized mobile phase system. For the pair of cys and his, the retardation factor of his is larger than that of cys on PVA plate while cys has larger retardation factor on Si-Gel plate. A similar trend was observed for trp/ala. The selectivity of tyr/thr and arg/asp was higher on the PVA plate, while phe/asp showed higher selectivity on the Si-Gel plate. The selectivity of met/gln on these two types of plates is similar. In summary, the selectivity of FITC-labeled amino acids is markedly different for most of the studied amino acids on PVA plates and Si-Gel plates even though both have hydroxyl functionalization.

2.3.3 *Efficiency*

The plate number (N) and plate height (H) describe the efficiency of separations TLC. In this work N was determined by PeakFit™ software, using statistical moment analysis of the chromatographic band and H was obtained with the equation H = N/L.

Different in the internal diameter of the capillary tubes used for the spotting the analyte solution on the ULTC plat can impact the sizes of the initial spots and

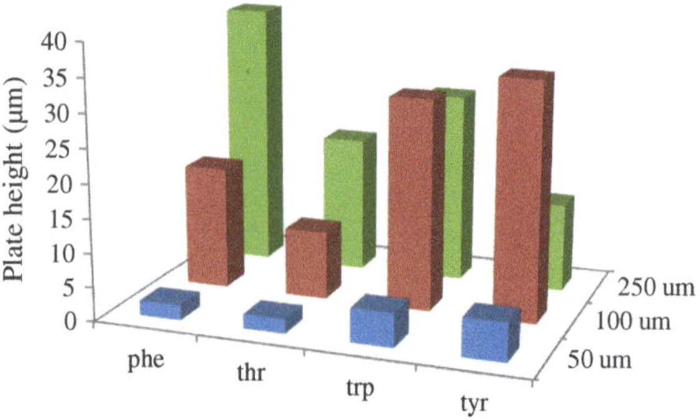

Fig. 2.11 Plate height of different capillary size

therefore the efficiency (Fig. 2.11). A larger i.d. capillary applies a larger original spot, which can lower the efficiency of the separation. Both 250 and 100 μm capillary showed lower efficiency than the 50 μm capillary. Therefore 50 μm capillary tube was selected for the remainder of the optimization studies.

The loading capacity of the three-layered plates was investigate using a 50 μm capillary tube and three-layer mat. Figure 2.12 shows that 7 mM solutions of analytes could be applied with minimal impact on the band dispersion. Significant band dispersion was noted for when 14 mM solutions were applied to the plate. The plate is clearly overloaded at this concentration. Comparing to a PVA-UTLC plate, 250 μm capillary tube and 7 mM solutions of analytes were used for the Si-Gel HPTLC plate without overloading. The HPTLC plate is much thicker than the UTLC plate. Therefore the loading capacity of HPTLC plate is larger. However if a 50 μm capillary tube was used for the HPTLC plate no fluorescent spot could be detected under UV light, which indicates that UTLC is a more sensitive technique than conventional TLC.

Sample volumes applied onto the plate also impact the efficiency (Figs. 2.13 and 2.14) with a 50 μm capillary tube and three-layered mat. For both 7 mM and 4 mM analyte concentration the smaller sample volume showed higher efficiency.

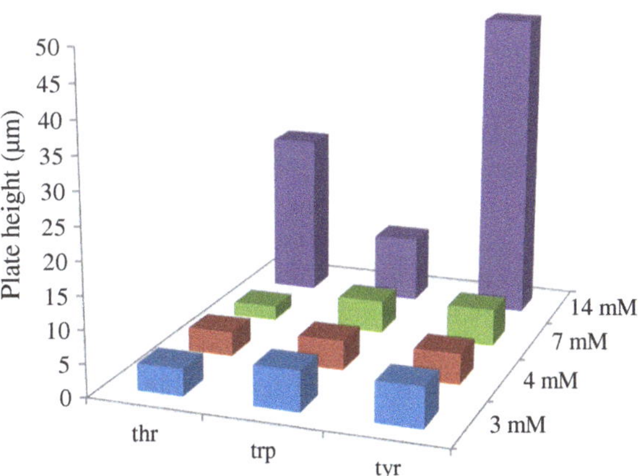

Fig. 2.12 Plate height of different analyte concentration

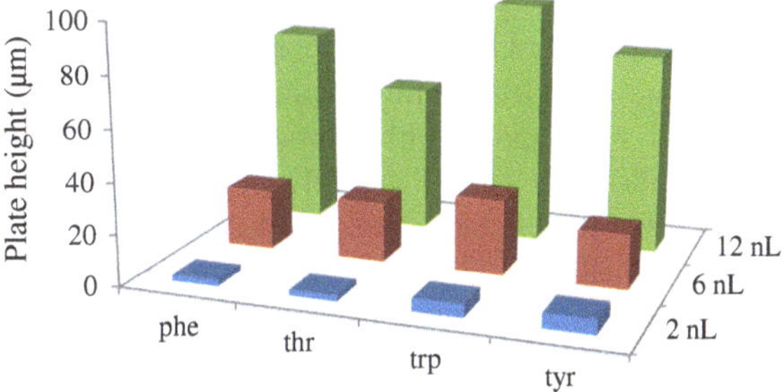

Fig. 2.13 Plate height of different analyte volume using 7 mM analyte

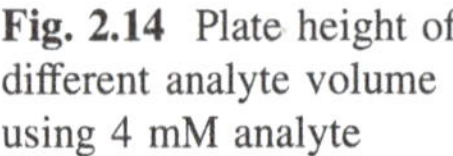

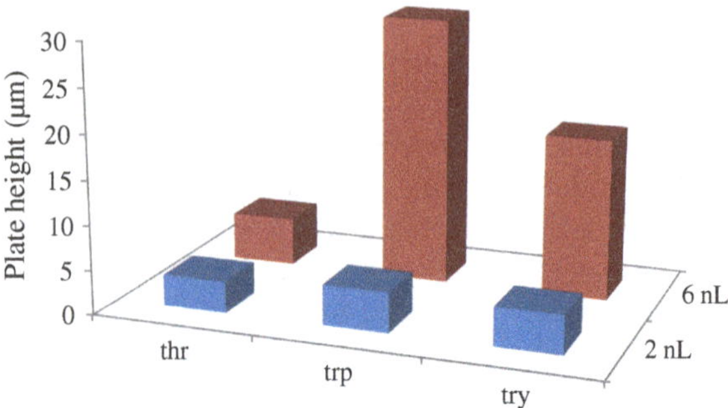

Fig. 2.14 Plate height of different analyte volume using 4 mM analyte

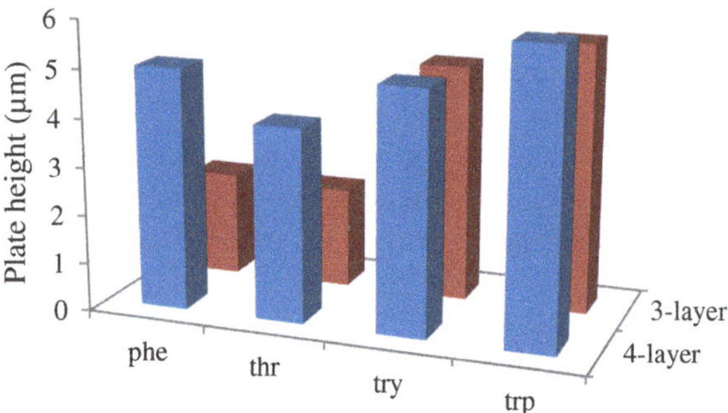

Fig. 2.15 Plate height of different mat thickness

Therefore, the minimum sample volume was applied each time to get the optimal efficiency for the currently used concentration range.

The mat thickness of the TLC plate can also impact the efficiency of the separation [25]. We studied the efficiency on PVA plates of different mat thickness (Fig. 2.15) with a 50 μm capillary tube and 7 mM solution. The one-layer and two-layered plates were overloaded because the mats are too thin as mentioned in Sect. 2.3.1. Also the application volume and analyte concentration used for this analysis are those optimized for the three layered thick mat. However, interestingly, it was observed that the efficiency on the four-layered thick plate is slightly lower than that on the three-layered thick plate. The reason could be the beads that form on the PVA plate as mentioned previously, may cause the heterogeneity of the plate, and therefore lower the efficiency.

The efficiency, plate number and plate height of the separation was compared with that on a Si-Gel HPTLC plate (Figs. 2.16 and 2.17). The efficiency on the PVA plate using the optimal condition mentioned above, 50 μm capillary tube, 3-layered mat, and 7 mM solution was studied. The distance traveled by the mobile phase on both PVA UTLC plate and Si-Gel plate is 3 cm. The result showed that the efficiency on the PVA plate is much higher than the efficiency on the Si-Gel HPTLC plate. The Si-Gel often has tailed bands when it is used to separate polar compounds, including amino acids. Therefore, PVA as a stationary phase provided an alternative for the separation of the polar compounds.

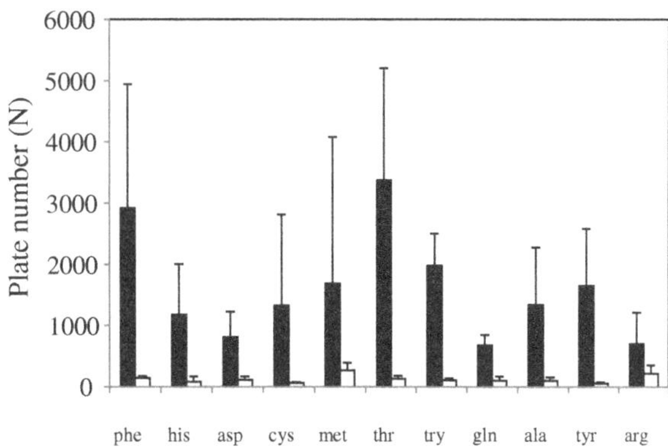

Fig. 2.16 Comparison of the plate number on PVA plate (*black*) and Si-gel HPTLC plate (*blank*). The error bars represent the standard deviation of three replicates

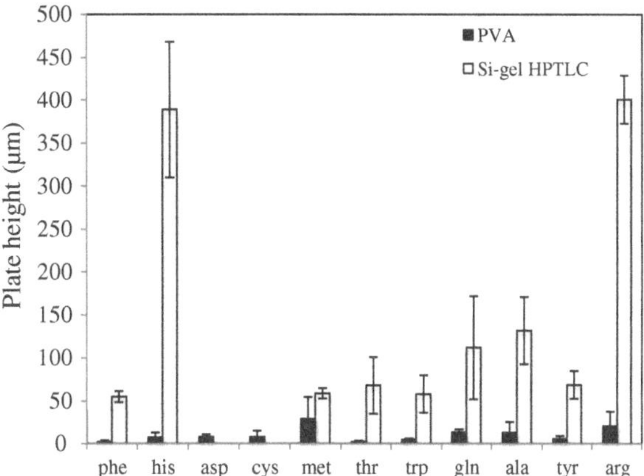

Fig. 2.17 Comparison of the plate height on PVA plate (*black*) and Si-gel HPTLC plate (*blank*). The error bars represent the standard deviation of three replicates

The plate height of asp and cys are not shown for Si-gel HPTLC due to significant tailing.

2.3.4 Separation of Amino Acids Using Ninhydrin as Visualization Reagent

Ninhydrin is the most widely-used detection agent for the visualization of amino acids by TLC. This visualization method was also studied with the electrospun PVA plates based on the optimized conditions for FITC-amino acid separations. Three-layered thick mats and a 50 μm i.d. capillary were therefore used. The ninhydrin reagent solution was sprayed onto the plate after development and the colored spots showed up after heating. No UV light or fluoresce detector are needed. The spots can be directly visualized after the ninhydrin reaction. The

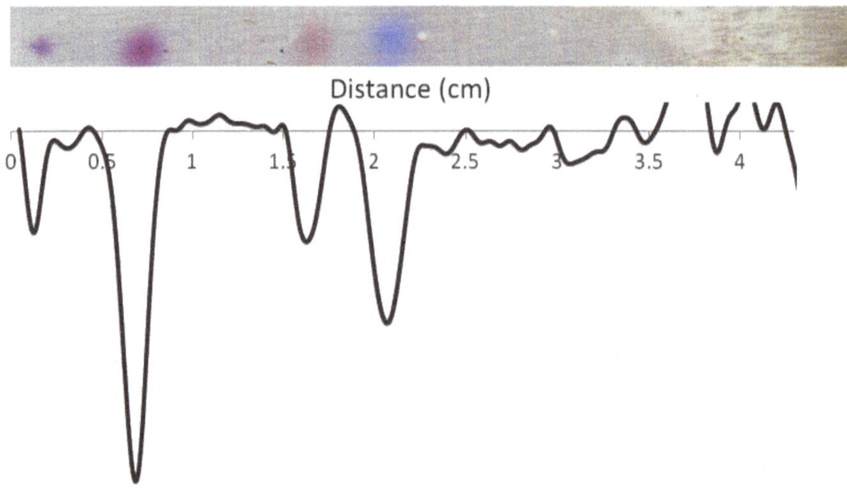

Fig. 2.18 Chromatogram of the separation of amino acids on electrospun PVA plate using ninhydrin as a visualization reagent

Table 2.2 hR_F, plate height, and color of spots of amino acids separated on PVA plate and plate height of conventional TLC using ninhydrin as a visualization reagent

	PVA UTLC			Literature HPTLC [30]
	hR_F	h (plate height/μm)	Color	h (plate height/μm)
Arg	1 ± 1	70 ± 40	Purple	83
Ala	14 ± 1	80 ± 10	Pink	96
Met	36 ± 1	30 ± 7	Orange	104
Phe	45 ± 1	30 ± 2	Blue	134

Fig. 2.19 Comparison of the selectivity of the FITC-labeled (*blank*) and unlabeled (*black*) amino acids on PVA plate

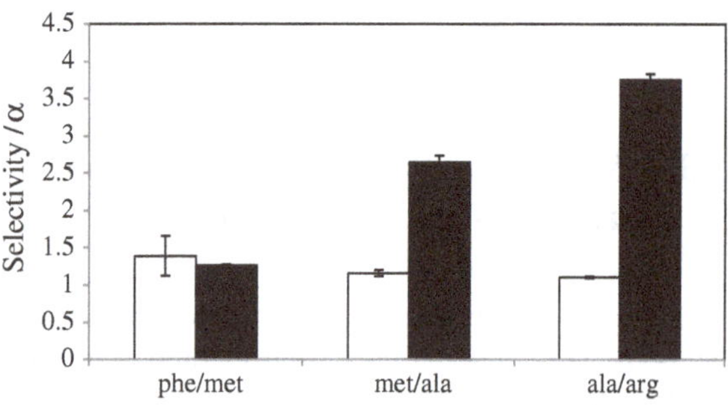

separation of four amino acids were shown in Fig. 2.18 and the hR_F and efficiency were calculated (Table 2.2).

The amino acids are arg, ala, met, and phe from left to right.

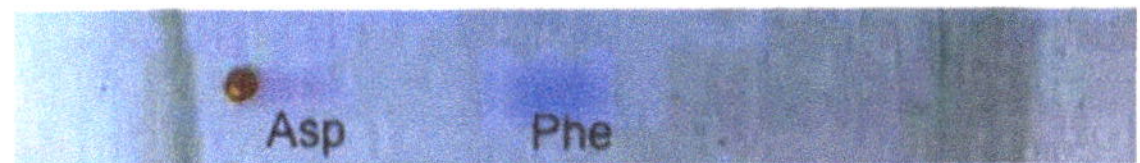

Fig. 2.20 A PVA-UTLC plate showing the separation of the hydrolysis products, asp and phe, of aspartame in diet coke in 2 cm

The mobile phase used to separate amino acids was optimized with butanol/ethyl acetate/water (5:5:1.5, volume ratio) showing the best selectivity. The amino acids without FITC labeling showed better selectivity compared to the FITC labeled amino acids on a PVA plate (Fig. 2.19). Four amino acids were baseline separated within 4 cm (Fig. 2.18) due to both improved selectivity and efficiency using a ternary mobile phase in composition of nontoxic organic solvents and water without any other additives. It is worthy to mention that the amino acids studied in this paper showed different colors (Table 2.2), which adds another dimension for the identification in addition to hR_F. Although the separation efficiency of amino acids is not as high as the FITC labeled amino acids on an electrospun PVA plate the efficiency of amino acids visualized by ninhydrin is higher than the efficiency of silica-gel monolithic UTLC that has been commercialized (80–100 μm) [29]. Flieger et al. [30] have reported high efficiency HPTLC separation of amino acids with ninhydrin as visualization reagent using salting-in TLC recently. The efficiency using PVA-UTLC and HPTLC [30] was compared in Table 2.2. The PVA-UTLC plate showed much higher efficiency for met and phe than that using HPTLC.

2.3.5 Analysis of Aspartame in Diet Coke

Analysis of aspartame, an artificial sweetener, and its breakdown products is important in the food industry since it is widely used in diet food and drinks as a sugar substituent. Change in pH or temperature can affect its stability and it degrades and forms asp, phe and methanol [31]. HPLC and TLC have been used for the analysis the breakdown products [32, 33]. In this work we used a PVA-UTLC plate to separate and identify the hydrolysis products, asp and phe. Diet coke was heated to reflux for one day to achieve a high yield of asp and phe [31]. The TLC mobile phase condition and ninhydrin treatment were similar to the method used in Sect. 2.3.4. The result is shown in Fig. 2.20. Asp and phe were well separated and could be readily identified from their color when compared to the color of their standards on PVA-UTLC plate after ninhydrin treatment. The colored ingredients in diet coke stayed in the origin and did not cause any inference. Therefore, a PVA-UTLC plate can be potentially used for aspartame analysis in food industry as a simple, fast, and low cost technique.

Butanol:ethyl acetate:water (5:5:1.5, volume ratio) was used as mobile phase.

2.4 Conclusions

A crosslinked PVA UTLC plate was fabricated and the fiber morphology was well maintained after soaked in water and the mobile phase used. The selectivity of FITC-labeled amino acids studied on a PVA plate was different from that on a Si-Gel plate except for met/gln. The efficiency depending on the mat thickness, capillary size, analyte concentration, and analyte volume was studied and optimized. For the three layered thick electrospun ULTC plates, smaller capillary tube and smaller analyte volume applied showed higher efficiency. Analyte concentrations below or equal to 7 mM showed similar efficiency, but the 14 mM analyte concentration showed much lower efficiency. The measured efficiency on the PVA ULTC plate was much higher than the efficiency on Si-gel plate. The separation of unlabeled amino acids showed higher selectivity and four amino acids were baseline separated using a simple and nontoxic ternary mobile phase. The hydrolysis products, asp and phe, of aspartame in diet coke were successfully separated and identified using a PVA-UTLC plate. This preliminary study of amino acid analysis also indicates the potential use of the PVA plate in the separations of other biomolecules.

References

1. Paradossi G, Cavalieri F, Chiessi E, Spagnoli C, Cowman MK (2003) J Mater Sci Mater Med 14:687
2. Dubey KA, Chaudhari R, Rao R, Bhardwaj YK, Goel NK, Sabharwal S (2010) J Appl Polym Sci 118:3490
3. Ichimura K (1996) Heterog Chem Rev 3:419
4. Brasch U, Burchard W (1996) Macromol Chem Phys 197:223
5. Yeom C, Lee K (1996) J Membr Sci 109:257
6. Zhang Y, Zhu PC, Edgren D (2010) J Polym Res 17:725
7. Han B, Li J, Chen C, Xu C, Wickramasinghe SR (2003) Trans IChemE 81:1385
8. Yang E, Qin X, Wang S (2008) Mater Lett 62:3555
9. Naebe M, Lin T, Staiger MP, Dai L, Wang X (2008) Nanotechnology 19:305702
10. Wang Y, Hsieh Y-L (2008) J Membr Sci 309:73
11. Tang C, Saquing CD, Harding JR, Khan SA (2010) Macromolecules 43:630
12. Kaspar J, Dettmer K (2009) Anal Bioanal Chem 393:445
13. Bhushan R, Reddy GP (1989) Biomed Chromatogr 3:233
14. Dale T, Court WE (1981) Chromatographia 14:617
15. Mazmierczak D, Ciesielski W, Zakrzewski RJ (2005) Planar Chromatogr 18:427
16. Norfolk E, Khan SH, Fried B, Sherma J (1994) J Liq Chromatogr 17:1317
17. Liu XD, Kubo T, Diao HY, Benjamas J, Yonemichi T, Nishi N (2009) Anal Bioanal Chem 393:67
18. Simon G, Liana G, Letitia G (2001) J Pharm Biomed Anal 26:681
19. Sherma J (2010) Anal Chem 82:4895
20. Mohammad A, Zehra A (2008) Proc Natl Acad Sci India 78A:11
21. Bhushan R, Reddy GP (1987) Anal Biochem 162:427
22. Mohammad A, Laeeq S (2007) J Planar Chromatogr Mod TLC 20:423

23. Cheng Y, Dovichi NJ (1988) Science 242:562
24. Maeda H, Ishida N, Kawauchi H, Tuzimura K (1969) J Biochem 65:777
25. Clark J, Olesik SV (2009) Anal Chem 81:4121
26. Vasta JD, Fried B, Sherma J (2010) J Liq Chromatogr Relat Technol 33:1028
27. Poole CF (2003) The essence of chromatography. Elsevier, Amsterdam
28. Clark J, Olesik SV (2010) J Chromatogr A 1217:4655
29. Poole SK, Poole CF (2011) J Chromatogr A 1218:2648
30. Flieger J, Tatarczak M (2008) J Chromatogr Sci 46:565
31. Homler BE (1984) Food Technol 38:50
32. Cheng C, Wu S (2011) J Chromatogr A 1218:2976
33. Conklin AR (1987) J Chem Educ 64:1065

Chapter 3
Homogeneous Carbon as Stationary Phase for Liquid Chromatography

Abstract Homogeneous edge-plane carbon stationary phase was prepared using silica as template. The carbon stationary phase was evaluated by the column efficiency, linear solvation energy relationships, and separation of polar compounds. Compared to other carbon phases, the edge-plane carbon showed unique selectivity, especially for polar analytes.

3.1 Introduction

Carbon stationary phases have been widely used in chromatographic research and applications [1, 2]. Carbon phases provide unique selectivity compared to chemically-bonded silica reversed-phase stationary phases [3, 4]. Moreover, research has shown that carbon phases can be used for the separation of structural isomers, due to the adsorption mechanism, and anions [5]. Additionally, unlike chemically-bonded silica gel stationary phases, carbon phases are stable under extreme pH conditions.

The current carbon phases, including Hypercarb [6], carbon-clad inorganic particles [7, 8] and low temperature glassy carbon coated particles [9], are composed of amorphous carbon that contains intertwining graphite ribbons without any ordered alignment. Therefore, there are at least two types of sites: edge-plane sites and basal-plane sites as introduced in Chap. 1 [10]. When the graphite layer is perpendicular to the surface, it is called an edge-plane site. When the graphite layer is parallel to the surface it is called a basal-plane site. Edge-plane sites are more polar than basal-plane sites. In addition, there may be functional groups attached to the edge-plane sites [11]. The presence of edge-plane and basal-plane sites can cause heterogeneity of the carbon surface and impact the chromatographic behavior.

The alignment of graphite layers can be controlled using a surface-directed liquid crystal assembly method [12]. AR meso phase (MP), a naphthalene polymer, has been used previously as a precursor for the synthesis of ordered carbon materials [12]. It can form ordered alignment at its liquid crystal state when heated to 300 °C. The anchoring state of the molecule depends on the nature of the substrates.

T. Lu, *Nanomaterials for Liquid Chromatography and Laser Desorption/Ionization Mass Spectrometry*, Springer Theses,
DOI: 10.1007/978-3-319-07749-9_3,

For example, it forms an edge-on anchoring state on alumina and silica, and a face-on anchoring state on platinum and silver [13]. The ordered alignment can be captured after the carbonization. When different substrates are used, edge-plane or basal-plane ordered carbon could be synthesized. The preparation of ordered edge-plane carbon nanorods has been reported by Hurt et al. [12]. However, the ordered carbon material for chromatographic application has never been reported.

Carbon stationary phases are of great interest in the separation of polar and ionic compounds. Hypercarb is also referred to as a "super reversed phase" because it provides stronger retention to analytes, especially polar solutes, compared to chemically-bonded silica stationary phases [14]. The main interactions between Hypercarb and solutes are induced dipole interactions [15]. Therefore, it is still used as a stationary phase of low polarity, which limits the application of Hypercarb for the analysis of more polar compounds. Recently, Lucy et al. reported using carboxylate modified Hypercarb to improve the separation of more polar analytes under the hydrophilic interaction liquid chromatography (HILIC) condition [16].

In this work, edge-plane carbon was prepared as a more polar stationary phase and used for the separation of polar and ionic analytes under reversed-phase condition. AR MP was used as a carbon precursor and porous silica gel particles were used as a support and a substrate that yields an edge-plane anchoring state. The interactions between the solutes and homogenous edge-plane carbon stationary phase were studied using linear solvation energy relationship (LSER). The separation of nucleotides and nucleosides, and amino acids and its derivative were performed using edge-plane carbon stationary phase under reversed-phase condition. The selectivity of edge-plane carbon was also compared with that of other carbon phases. The efficiency of the edge-plane carbon stationary phase was studied as a function of linear velocity, and the column batch-to-batch reproducibility was comparable to that found for commercial stationary phases.

3.2 Experiment

3.2.1 Materials

AR MP pitch was obtained from Mitsubishi Gas Chemical America (New York, NY). Forming gas (5 % hydrogen/95 % nitrogen, standard grade) was purchased from Praxair. Ammonium acetate, mono potassium phosphate, sodium nitrate, dichloromethane and acetonitrile (ACN) were purchased from Fisher Scientific. In-house nanopure water was used for chromatography study. Acrylamide, indapamide, N-isopropylacrylamide, DMF, trifluoricacetic acid (TFA) and all the analytes for LSER study were purchased from Sigma Aldrich (St. Louis, MO). Cytidine (C), uridine (U), cytidine 5′-monophosphate (CMP), cytidine 5′-diphosphocholine (CDP), cytidine 5′-triphosphate (CTP), histidine (His), phenylalanine (Phe), tryptophan (Trp), tyrosine (Tyr), and fluorotryptophan were purchased from Sigma Aldrich (St. Louis, MO). Silica gel particles were from Phenomenex (Torrance, CA).

3.2.2 Edge-Plane Carbon Stationary Phase Preparation

The method for coating of the AR MP on to the silica gel particles used in this work was similar to the procedure reported by our group previously [9]. Silica gel particles were placed in the oven (150 °C) to remove the water adsorbed before using. Then the silica gel particles and 0.1 % AR MP solution (wt%, in dichloromethane) were added into a fluidized bed. The amount of silica gel particles and AR MP was calculated to achieve 5, 25, 50 and 80 % carbon load percentage (wt%). The carbon load was calculated using the following equation:

$$\text{Carbon load} = [\text{mass(carbon)}/\text{mass(silica)}] \times 100\,\%. \tag{3.1}$$

A flow of nitrogen was introduced into the fluidized bed from the bottom to form a homogenous suspension and to assist in the evaporation of the solvent. The fluidized bed was placed in a water bath that was held at room temperature because otherwise solvent evaporation would have caused a temperature drop in the suspension. After the solvent was completely removed the particles were transferred into a tube furnace (Lindberg/Blue M Asheville NC, Model: STF55346C-1). The particles were pyrolyzed under a forming gas atmosphere. For the preparation of edge-plane carbon the particles were first heated to 300 °C in 10 min and then the temperature was held at 300 °C for four hours to allow the AR meso phase to form ordered alignment. After that the temperature was slowly increased to 700 °C in two hours and held at 700 °C for 1 h. The final temperature was set to 700 °C because the same pyrolysis temperature has also been used in the preparation of other types of carbon phases [8]. Then the furnace was turned off and allowed to cool to room temperature. For the preparation of amorphous carbon, only the heating protocol was changed. The temperature was increased directly to 700 °C with a ramp rate of 2 °C/min and the temperature was held at 700 °C for 1 h [12]. Without holding the temperature at 300 °C AR MP could not form ordered alignment.

3.2.3 Characterizations of Carbon Coated Particles

Transmission electron microscopy. The carbon particles were dispersed in methanol in a sonicator. Then a drop of the suspension was placed on to a lacey carbon copper grid for transmission electron microscopy (TEM) (Electron Microscopy Sciences, Hatfield, PA). High-resolution TEM was performed using a Tecnai F20 system. Scanning transmission electron microscopy (STEM) was also performed using the same instrument. High-resolution TEM was used to characterize the alignment of the graphite layers. STEM, which records the image from the secondary electrons that are scattered by the sample surface, was used to visualize the core-shell structure of the carbon coated silica beads. Porous silica gel particles were used for the preparation of edge-plane carbon. However, when the particles were observed in TEM instrument the porous structure was not stable and decomposed rapidly. Therefore, it was very difficult to focus the image before its decomposition. Nonporous silica gel particles (200 nm diameter, Fiber Optics, New Bedford, MA) were coated following the same procedure and used for the TEM and STEM studies.

Scanning electron microscopy. The micrographic images were obtained with a Hitachi S-4300 (Hitachi High Technologies, America, Inc., Pleasanton, CA) scanning electron microscope. Before the SEM analysis the sample was sputter coated with gold for 2 min at 10 μA.

X-ray photoelectron spectroscopy. The X-ray photoelectron spectroscopy (XPS) was performed using a Kratos Axis Ultra XPS. Mg X-ray gun was used for the analysis.

Thermogravimetric Analysis. A TGA Q50 (PerkinElmer) was used for thermogravimetric analysis (TGA). A small amount of sample, ~5 mg, was used each time. The temperature was increased to 1,000 °C at 10 °C/min under air. The amount of carbon load was calculated by subtracting the final weight of silica particles from the weight at ~100 °C. At this temperature the water adsorbed evaporated and only carbon and silica gel remained.

3.2.4 Chromatography Studies

The edge-plane carbon coated silica particles, amorphous carbon coated silica particles, Hypercarb, and bare silica gel particles were packed into fused silica tubings (250 μm id, 350 μm od, 35 cm long) using a previously reported method [9]. Briefly, the particles were first dispersed in acetonitrile in a sonicator for 20 min. Then the suspension was placed in a small reservoir, 4 mm id × 10 cm long stainless steel tubing. The suspension solution was pushed into the fused silica tubing using a syringe pump (ISCO LC-2600 precision syringe pump, ISCO, Lincoln, NE) filled with acetonitrile. The packing was held in the column with a microbore column end fitting (U-434, Upchurch Scientific, Inc., Oak Harbor, WA). The pressure was slowly increased to 3,500 psi and then the pressure was held for about 4 h. After that the pump was turned off and the pressure was allowed to be released through the end of the column slowly.

An ISCO LC-2600 syringe pump was used to deliver the mobile phases for the LC separation. All of the chromatographic studies were performed under constant pressure condition since a previous study showed that the flow in a micro column is more stable under constant pressure than that under constant flow rate [17]. The samples were injected using a 60-nL high-pressure injection valve (Valco Instruments, Houston, TX). A UV-1000 detector (Thermo Scientific) was used with a fused silica tubing (100 μm id, 200 μm od, polyimide coating removed) as a flow cell in the detector. Therefore, the path length of the detector was 100 μm, and the illuminated volume was 15 nL.

Amides were chosen in this work because they are highly polar compounds. The carbon coverage was determined by the separation of amides mixture. The details were discussed in Sect. 3.3.2. The amides samples were dissolved in acetonitrile (1 mg/mL). All the analytes for LSER were dissolved in acetonitrile (3 mg/mL) with acetone that was used for the determination of dead time. The nucleosides and nucleotides were dissolved in 50 mM ammonium acetate buffer solution (1 mg/mL). The amino acids were dissolved in 50 mM formate buffer solution (1 mg/mL).

3.3 Results and Discussion

3.3.1 Synthesis and Characterization of Edge-Plane Carbon Coated Particles

The edge-plane carbon particles were prepared using silica gel particles as a support. Carbon-coated silica gel particles have been reported [7, 8]. However, the alignment of carbon was not controlled. In this work, by using AR MP as a carbon precursor, homogenous edge-plane carbon coated silica gel particles were prepared. The AR MP precursor was coated on to the surface of silica gel particles by weak interactions in solution following a procedure that has been used to prepare low temperature glassy carbon [9]. Although in previous work AR MP was dissolved in pyridine before use [12], in this work dichloromethane was used as solvent. The reason is that pyridine is very difficult to remove in the fluidized bed even under higher temperature (80 °C). We found that dichloromethane is also a good solvent for AR MP and the removal of dichloromethane in the fluidized bed is much faster. The carbonization was performed following a previous reported temperature program [12].

The formation of the edge-plane carbon coating was illustrated by TEM and STEM studies. The high resolution TEM image (Fig. 3.1) shows that the edge-plane carbon alignment was formed on the silica gel beads. The STEM image (Fig. 3.2) illustrates that the carbon coated silica gel beads have a characteristic core-shell structure, which further demonstrates the successful coating on the silica beads. The thickness of the carbon coating, determined from the STEM, is 12 ± 3 nm. Therefore, the edge-on anchoring state can be aligned in a thickness of at least 12 nm on the silica gel beads. It has been reported that the thickness of the alignment can be up to 1 μm [18].

The bulk homogeneity of the carbon coated silica gel particles was studied using XPS. Hypercarb and amorphous carbon coated silica gel were also studied for comparison purpose. The XPS of edge-plane and amorphous carbon-coated silica gel particles showed strong Si signal (Figs. 3.3 and 3.4). Because the carbon coating was thin the X-ray can penetrate the carbon layer and generate signals

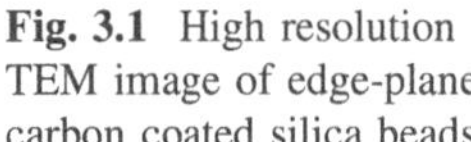

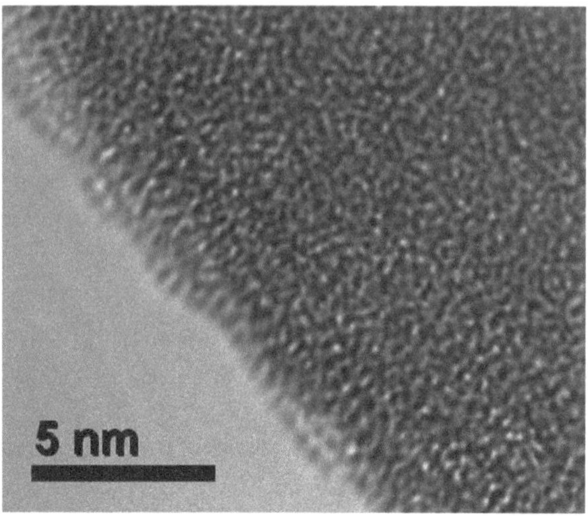

Fig. 3.1 High resolution TEM image of edge-plane carbon coated silica beads

Fig. 3.2 STEM image of edge-plane carbon coated silica beads

from the silica support. Because the silica gel template was removed from Hypercarb, no Si signal was observed for Hypercarb (Fig. 3.5).

The peak width has been used to compare the homogeneity of the alignment of graphite layers [19]. The carbon material with heterogeneous graphite layers would show a broader peak because the electronic environment of the carbon atoms is different. Table 3.1 shows the full width at half maximum (FWHM) of

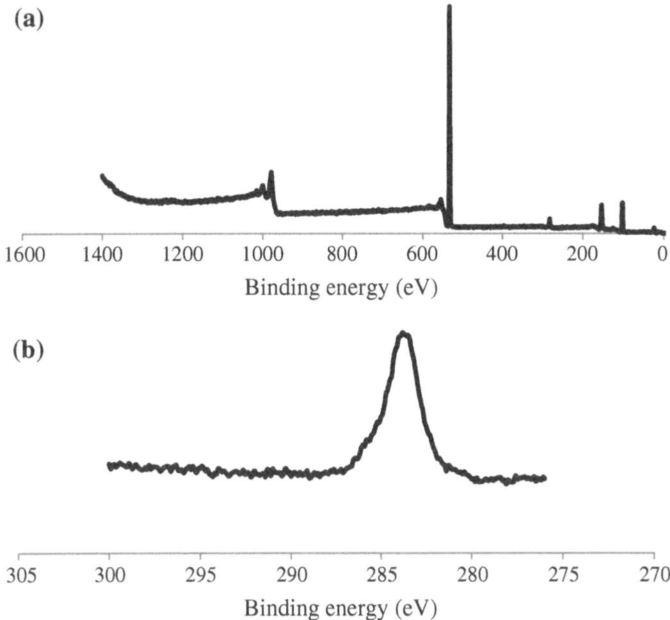

Fig. 3.3 XPS spectrum of edge-plane carbon: **a** survey, **b** carbon peak

different carbon particles for HPLC packings. Amorphous carbon shows higher FWHM than that of edge-plane carbon. This indicates that the homogeneity of the graphite layers of amorphous carbon is lower than that of edge-plane carbon. Hypercarb also showed a smaller FWHM compared to edge-plane carbon. The reason may be that Hypercarb is considered to be composed of a high degree of basal-plane carbon with small amount of edge-plane sites. Another possible reason is that the edge-plane carbon may contain organic functionalities, which could also cause a broader peak. In addition, a small $\pi - \pi^{*}$ peak at $\sim$287 eV was observed for Hypercarb. This peak is due to the largely conjugated basal-plane sites [19]. However, this peak was not observed for edge-plane and amorphous carbon. There are two possible reasons: (1) edge-plane carbon and amorphous carbon don't have largely conjugated basal-plane sites; (2) the amount of the conjugated basal-plane is too low and XPS cannot detect them. The FWHM of amorphous carbon is similar compared to the carbon-clad alumina and zirconia particles that are prepared by CVD method that were discussed in Chap. 1 [8]. It is expected because the alignment of the carbon particles is not controlled using the CVD method. In addition, the final pyrolysis temperatures of the two methods are same [8].

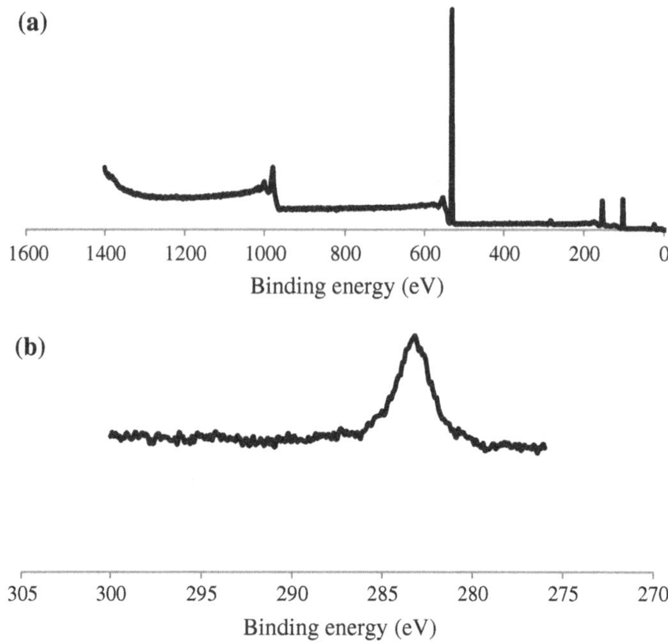

Fig. 3.4 XPS spectra of amorphous carbon: **a** survey, **b** carbon peak

3.3.2 Optimization of Carbon Load

The amount of carbon coated on the silica gel particles is important to achieve the optimal chromatographic behavior. If the amount of carbon is too small the surface of the silica gel could not be completely covered by carbon. The bare silica surface can contribute to the retention and separation, especially when polar compounds are separated. The surface coverage was studied using a mixture of amides compounds. The mixture was separated on the bare silica and carbon phases with different carbon load under a HILIC condition and the selectivity was compared. The structures of amides are shown in Fig. 3.6. Amides are highly polar compounds and they can be well separated using a bare silica gel column (Fig. 3.7). When 5 and 25 % carbon coated particles were used, they showed similar selectivity to the separation on the bare silica gel (Figs. 3.8 and 3.9). However, the bands are much narrower than those obtained from bare silica and retention time decreased. When the carbon load was increased to 50 %, different selectivity was observed (Fig. 3.10): the elution order of 1 and 2 and are reversed, and 1 and 3 co-eluted. The difference in selectivity is believed to be a result of the increased carbon coverage. This is a strong evidence of high coverage of edge-plane carbon on the silica gel particles. From theoretical calculation similar to a previous

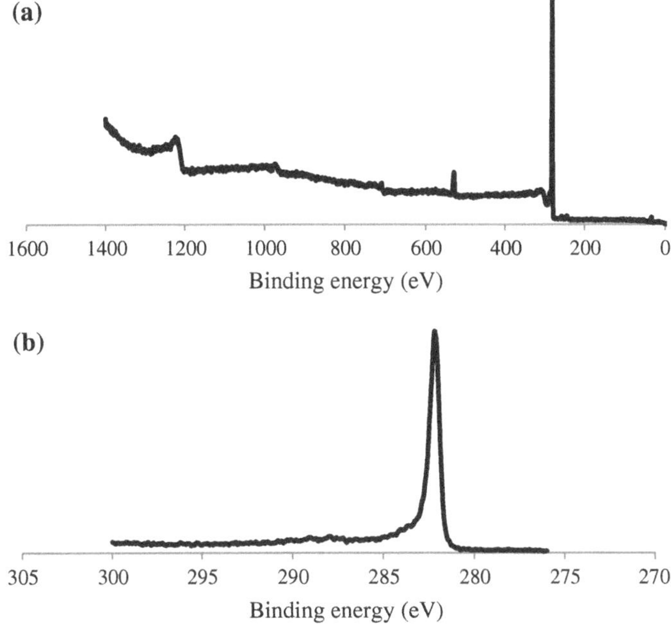

Fig. 3.5 XPS spectra of Hypercarb: **a** survey, **b** carbon peak

Table 3.1 FWHM of XPS peaks from edge-plane carbon, Hypercarb, and amorphous carbon

	FWHM (eV)
Edge-plane carbon	2.08
Amorphous carbon	2.28
Hypercarb	0.61
Carbon-clad alumina/zirconia [8]	2.3

reported method [8], 50 % carbon load forms a monolayer of carbon coating on the silica gel particles. It is in good agreement with the experiment result.

However, when the carbon load was further increased to 80 %, agglomeration was observed (Fig. 3.11).

3.3.3 Van Deemter Plot

Van Deemter equation is used to describe the band broadening in chromatographic systems. As described in Chap. 1, when the plate height, H, is plotted against the linear velocity, u, A, B and C terms can be determined by fitting the curve using

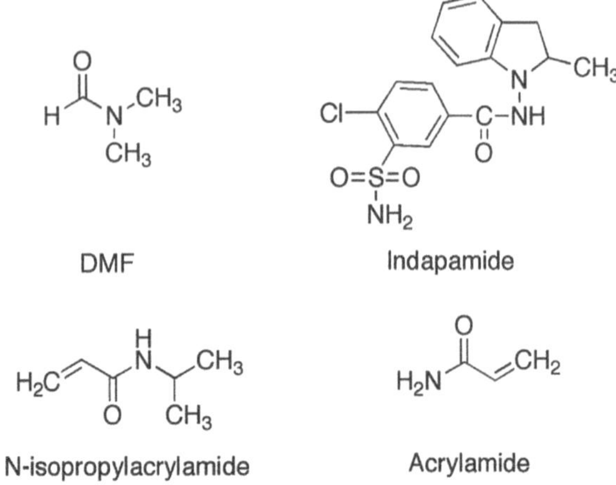

Fig. 3.6 Structures of amides

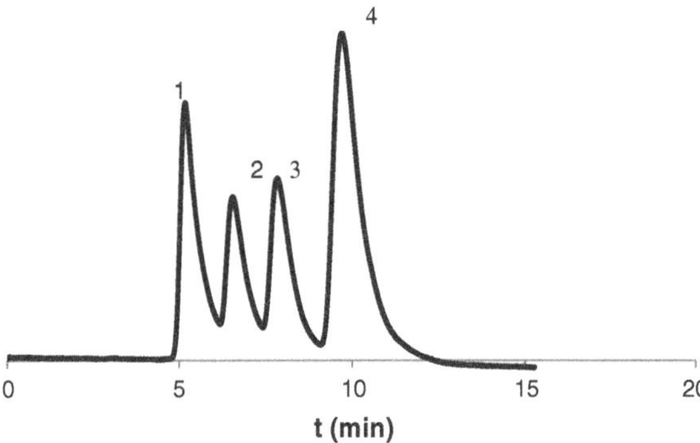

Fig. 3.7 HILIC separation of acrylamide, indapamide, N-isopropylacrylamide and DMF using bare silica gel. *1* indapamide, *2* N-isopropylacrylamide, *3* acrylamide, *4* DMF. Mobile phase: 95 % acetonitrile and 5 mM KH_2PO_4 buffer. Flow rate: 5 μL/min. UV detection at 214 nm

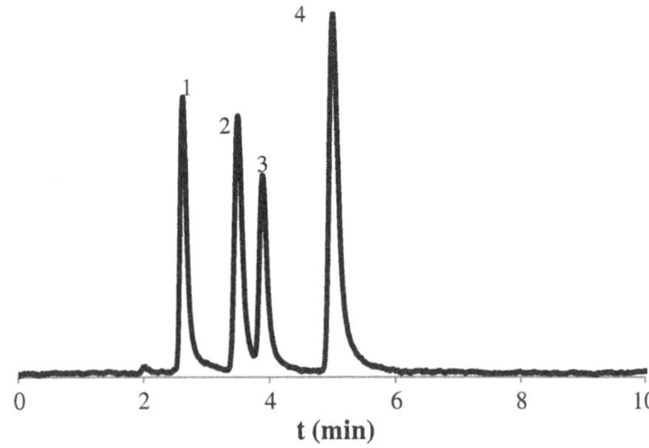

Fig. 3.8 HILIC separation of acrylamide, indapamide, N-isopropylacrylamide and DMF using 5 % carbon load edge-plane carbon as stationary phases. *1* indapamide, *2* N-isopropylacrylamide, *3* acrylamide, *4* DMF. Mobile phase: 95 % acetonitrile and 5 mM KH_2PO_4 buffer. Flow rate: 5 μL/min. UV detection at 214 nm

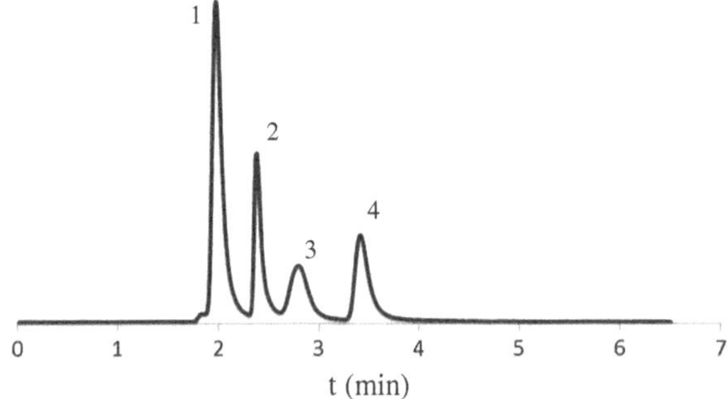

Fig. 3.9 HILIC separation of acrylamide, indapamide, N-isopropylacrylamide and DMF using 25 % carbon load edge-plane carbon as stationary phases. *1* indapamide, *2* N-isopropylacrylamide, *3* acrylamide, *4* DMF. Mobile phase: 95 % acetonitrile and 5 mM KH_2PO_4 buffer. Flow rate: 5 μL/min. UV detection at 214 nm

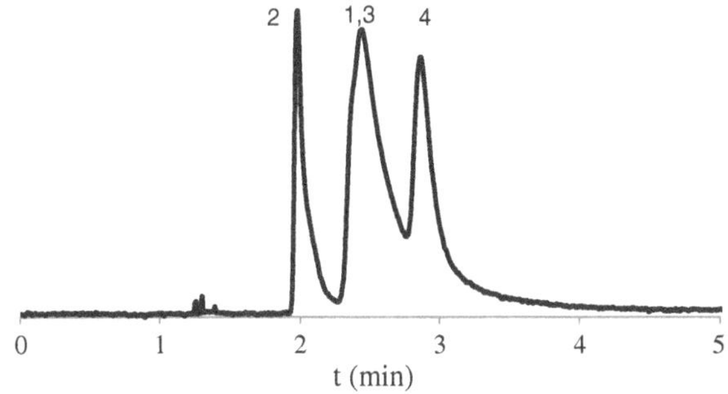

Fig. 3.10 HILIC separation of acrylamide, indapamide, N-isopropylacrylamide and DMF using 50 % carbon load edge-plane carbon as stationary phases. *1* indapamide, *2* N-isopropylacrylamide, *3* acrylamide, *4* DMF. Mobile phase: 95 % acetonitrile and 5 mM KH_2PO_4 buffer. Flow rate: 5 μL/min. UV detection at 214 nm

Fig. 3.11 SEM image of 80 % carbon load edge-plane carbon coated silica gel particles

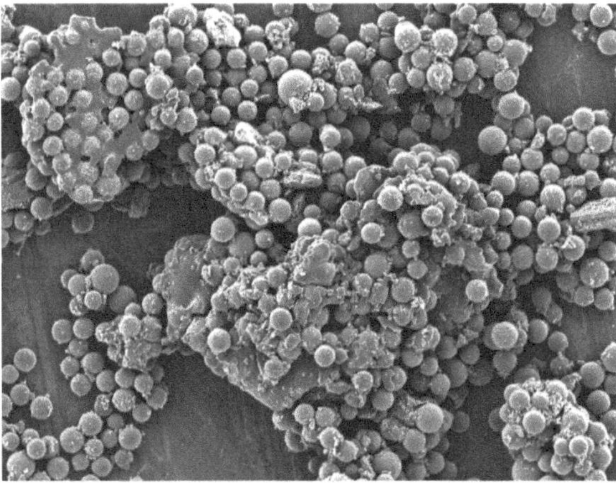

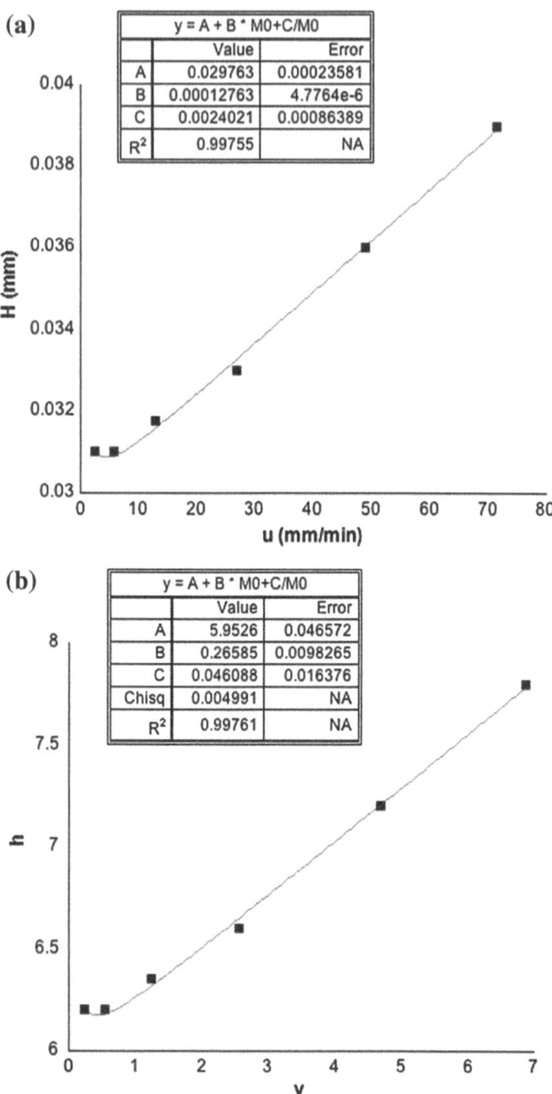

Fig. 3.12 Van Deemter plot of edge-plane carbon column: **a** H versus u **b** h versus v. Mobile phase: 60 % ACN in water. Phe was used as the solute. Acetone was used as a dead time marker in order to determine the linear velocity

van Deemter equation. Van Deemter plot and the curve fitting result were obtained here using phenylalanine as a solute and 60 % ACN in water as a mobile phase (Fig. 3.12a).

The minimum plate height, H_{min}, is of great interest because it represents the best efficiency of the column. H_{min} of the edge-plane carbon capillary column was determined using the following equation:

$$H_{min} = A + 2(B \times C)^{0.5}. \tag{3.2}$$

From the constants that were obtained from curve fitting, H_{min} is determined to be 31 μm. In order to compare the efficiency of stationary phases that are packed with different particle sizes, reduced plate height, $h = H/d_p$, can used. The reduced plate height is unitless and does not relate to the particle size. The minimum reduced plate height, h_{min}, for the edge-plane carbon column is 6. It has been reported that the h_{min} of carbon-clad zirconia ($d_p = 3$ μm) is <3 [20]. The lower efficiency may due to the interaction with the free silanol groups that are not covered by carbon. Another possible reason is the packing condition of the column. Improvement of the packing condition may increase the efficiency.

Figure 3.12 b shows the plot of the reduced plate height, h, versus the reduced velocity, $v = ud_p/D_m$. The diffusion coefficient was calculated using the Wilke-Chang equation. The reduced velocity is also unitless and does not related to the particle size, solute or mobile phase. Van Deemter equation was used to fit the curve. A term is larger compared to the carbon-clad zirconia [20]. Similar to h_{min}, it may due to the coverage of the silanols and the packing of the column. B term is smaller compared to the carbon-clad zirconia [20]. The reason may be the fitting of B term is not accurate. In order to determine B term more accurately, the efficiency at very low flow rate is needed. The value of C term is comparable with the carbon-clad zirconia column [20].

3.3.4 Linear Solvation Energy Relationship

LSER is widely-used to study the intermolecular interactions between the analytes and the stationary phases [21]. It is a useful tool to investigate the surface property of stationary phases. Equation (3.3) has been used for LSER study for liquid chromatography [21]:

$$Log\ k = C + m\ V_x + r\ R_2 + s\ \pi_2^* + a\ \Sigma\alpha_2^H + b\ \beta_2^H, \tag{3.3}$$

where k is the retention factor. The solute descriptors include the solute molecular volume V_x, calculated using McGowan's algorithm, excess molar refraction R_2, dipolarity/polarizability π_2^*, and hydrogen bond acidity and basicity $\Sigma\alpha_2^H$ and $\Sigma\beta_2^H$. The fitting coefficients are called system constants. Each is related to the corresponding interactions from the specific system, e.g. the stationary phase and the mobile phase. When the mobile phase used is the same, the system constants are characteristic to the stationary phase.

LSER has been used to study the intermolecular interactions on Hypercarb and the carbon phases prepared using CVD method by Carr et al. [5]. In their work, an LSER model without R_2 term, showed the best fitting results for the carbon phases (Eq. 3.4) [5].

Table 3.2 Solutes descriptors for LSER study

	V_x	R_2	π_2^*	α_2^H	β_2^H
Toluene	0.857	0.601	0.52	0	0.14
o-xylene	0.998	0.663	0.56	0	0.16
Benzene	0.716	0.61	0.52	0	0.14
Propiophenone	1.155	0.804	0.95	0	0.51
Ethylbenzene	0.998	0.613	0.51	0	0.15
3-phenol-1-propanol	1.198	0.821	0.9	0.3	0.67
Fluorobenzene	0.7341	0.477	0.57	0	0.1
Chlorobenzene	0.8388	0.718	0.65	0	0.07
p-xylene	0.9982	0.613	0.52	0	0.16
2-isopropylphenol	1.1978	0.82	0.84	0.57	0.36
Benzonitrile	0.8711	0.742	1.11	0	0.33
Phenetole	1.057	0.681	0.7	0	0.32
3-methylbenzyl alcohol	1.057	0.815	0.9	0.33	0.59
4-methylbenzyl alcohol	1.057	0.81	0.88	0.33	0.6
m-xylene	0.9982	0.623	0.52	0	0.16
Phenol	0.7751	0.805	0.89	0.6	0.3
Acetophenone	1.014	0.818	1.01	0	0.48
4-ethylphenol	1.057	0.8	0.9	0.55	0.36
4-bromophenol	0.95	1.08	1.17	0.67	0.2
Benzyl alcohol	0.916	0.803	0.87	0.33	0.56
Styrene	0.955	0.849	0.65	0	0.16
1,4-dichlorobenzene	0.962	0.825	0.75	0	0.02
Butyrophenone	1.2957	0.797	0.95	0	0.51
1,2-dibromobenzene	1.006	1.19	0.96	0	0.46
Methylbenzoate	1.073	0.733	0.85	0	0.46

The solute descriptors were from literature references [17, 22]

$$Log\ k = C + m\ V_x + s\ \pi_2^* + a\ \Sigma\alpha_2^H + b\ \beta_2^H. \tag{3.4}$$

An LSER study was performed for the edge-plane carbon phase to investigate the interactions between the analytes and the edge-plane sites. The solute descriptors of the analytes used are listed in Table 3.2. Similar to the previous results [5], Eq. (3.4) provided better fitting results for the edge-plane carbon phase. Therefore, LSER model was examined using Eq. (3.4) in this work.

The system constants obtained is listed in Table 3.3. The value of r is 0.962, which is comparable to the values that have been reported for other carbon phases [5]. Two solutes were outliers, including 4-bromophenol and benzyl alcohol. The cross correlation between the variables is listed in Table 3.4. Two correlation coefficients are relative large, >0.5, including V_x/β_2^H and π_2^*/β_2^H. However, removal of β_2^H term yields poor fitting result. Large correlation coefficients have also been observed in other LSER studies including those using carbon phases [5, 21]. The large correlation coefficients may be due to the aromatic compounds

Table 3.3 System constants from LSER regression results for edge-plane carbon phase

	Coefficients	Standard error
c	−1.0	0.3
m	1.3	0.3
s	0.7	0.2
a	−2.3	0.2
b	−0.6	0.3

$r = 0.962$

Conditions 60:40 acetonitrile/water mobile phase. 5 μL/min, 210 nm detection

Table 3.4 Cross-correlation matrix r^2

	V_x	π_2^*	α_2^H	β_2^H
V_x	1			
π_2^*	0.428	1		
α_2^H	0.178	0.354	1	
β_2^H	0.631	0.641	0.42	1

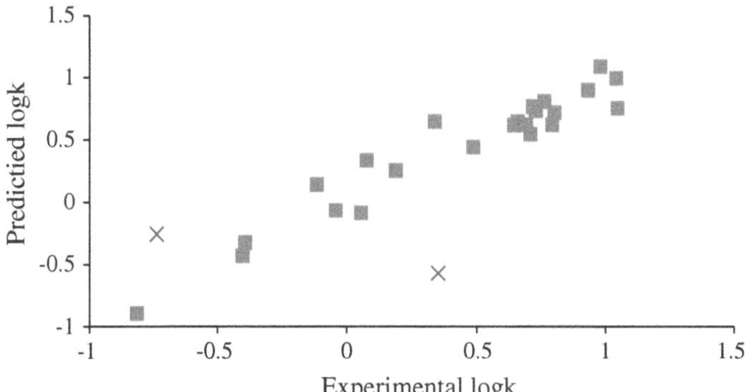

Fig. 3.13 Plot of log k based on the LSER versus log $k_{exp.}$ on edge-plane carbon stationary phase. Conditions: 60:40 acetonitrile/water mobile phase. 5 μL/min, 210 nm detection. Two outliers were marked as ×

that were used as solutes [21]. Figure 3.13 shows the plot of the predicted logk against the experimental log k. Two outliers were marked in the plot as ×.

The LSER results from the edge-plane carbon were compared to the results from other carbon phases. Hypercarb and carbon particles prepared using CVD method showed similar trend when examined using LSER model with 65 % ACN in water as mobile phase [21, 23]. The mobile phase for the LSER study of the edge-plane carbon phase was 60 % ACN in water. The composition of the mobile phase is very similar to the reported mobile phase condition, only 5 % difference, so it should not cause significant difference in retention. The difference in the system constants should be due to the stationary phases.

Edge-plane carbon showed different m coefficient compared to amorphous carbon phases. The carbon particles prepared by CVD method is amorphous carbon because the carbon alignment is not controlled. The amorphous carbon phases show larger m coefficient compared to chemically bonded reversed-phase stationary phase and thereby, they are considered to be more "hydrophobic" [5]. However, for edge-plane carbon phase, m coefficient is smaller than the other carbon phases, 1.3 versus 1.7–2. From the value of m coefficient we can conclude that the edge-plane carbon phase is less "hydrophobic". It is expected because that the edge-plane sites are more polar. Additionally, there may be hydrophilic functionalities attached to the edge-plane sites. Both of them can decrease the hydrophobicity of the surface. The s coefficient of edge-plane carbon is very close to that of the other carbon phases [5].

The hydrogen bonding interaction on edge-plane carbon is different compared to the amorphous carbon phases. For the amorphous carbon phases, b coefficient is a more significant contributor of the decreased retention than a [5]. However, for the edge-plane carbon a significant different trend was observed. The major contribution for the decreased retention is from a instead of b. The reason may be that the organic functionalities can donate protons and interact with the basic compounds. As a result, the basic solutes can be more retained on the stationary phase and thereby, the contribution in the decrease of the retention from b coefficient is smaller.

Therefore, based on the LSER model, we can conclude that the edge-plane carbon showed less hydrophobicity and significantly different hydrogen bonding interactions compared to the amorphous carbon phases. This could lead to different selectivity using the edge-plane carbon as a stationary phase for liquid chromatographic separations.

3.3.5 Separation of Nucleosides and Nucleotides

The separation of nucleosides and nucleotides is of great importance in pharmaceutical industry. Separations using gradient elution on Hypercarb column [24], ion-pair chromatography [25], HILIC [26, 27] and capillary electrochromatography [28] have been reported. Direct separation under reversed-phase conditions is challenging due to the highly polar and the ionic nature of these compounds. The structures of the nucleosides and nucleotides used are shown in Fig. 3.14.

In this work a mixture of nucleosides and nucleotides was separated using the edge-plane carbon stationary phase under an isocratic reversed-phase condition (Fig. 3.15). Because they are very polar analytes, especially for nucleotides, mobile phase (5 % acetonitrile and 95 % 50 mM ammonium acetate buffer) with low eluent strength was used. The elution order is typical for a reversed-phase separation. The most polar nucleotide, CTP, eluted first. The CMP eluted before CDP because CDP has a hydrophobic chain which has strong interactions with the stationary phase (Fig. 3.15). The less polar analytes, uridine and cytidine, retained

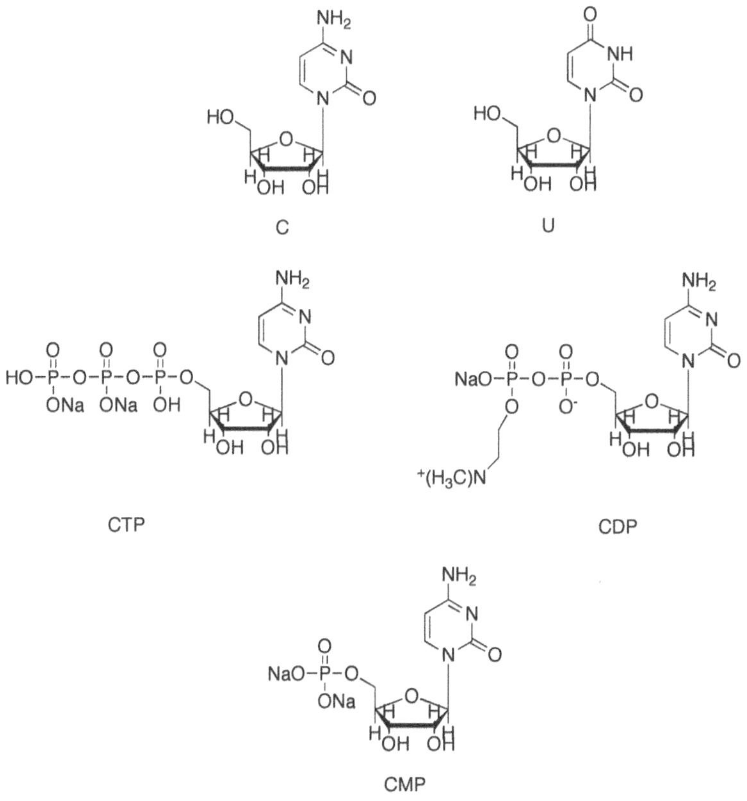

Fig. 3.14 Structures of CMP, CDP, CTP, U and C

on the column for longer time. The same mixture was also separated using the amorphous carbon as a stationary phase under an optimized condition (Fig. 3.16). Even when a mobile phase with very weak eluent strength (5 % acetonitrile and 95 % 50 mM ammonium acetate buffer) was used the mixture was not separated. CTP and CMP coeluted. Uridine and cytidine were not resolved either. The only difference between the amorphous carbon and the edge-plane carbon is the alignment of the carbon, i.e. whether the basal-plane sites or the edge-plane sites interact with the analytes. The separation result shows the importance of controlling the alignment of the carbon for the best separation. The result was compared to Hypercarb as well. Figure 3.17 shows the separation using Hypercarb as a stationary phase under an optimized condition. The nucleotides were poorly resolved. This may due to the significant amount of basal-plane sites on Hypercarb. As mentioned, the basal-plane sites are less polar and therefore they have low selectivity to the highly polar nucleotides. Another difference is that the elution order of cytidine and uridine is reversed on Hypercarb compared to that on edge-plane carbon. This is more evidence that the selectivity of the edge-plane carbon is unique and different from other carbon materials.

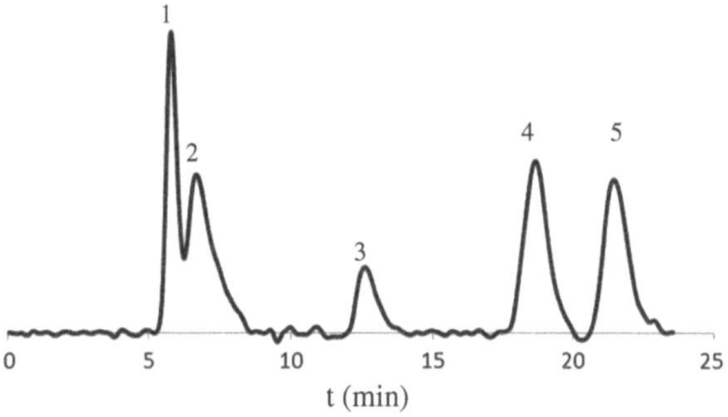

Fig. 3.15 Separation of nucleotides and nucleosides using edge-plane carbon as stationary phase. *1* CTP, *2* CMP, *3* CDP, *4* U, *5* C. Mobile phase: 5 % acetonitrile and 50 mM ammonium acetate buffer. Flow rate: 4 μL/min. UV detection at 254 nm

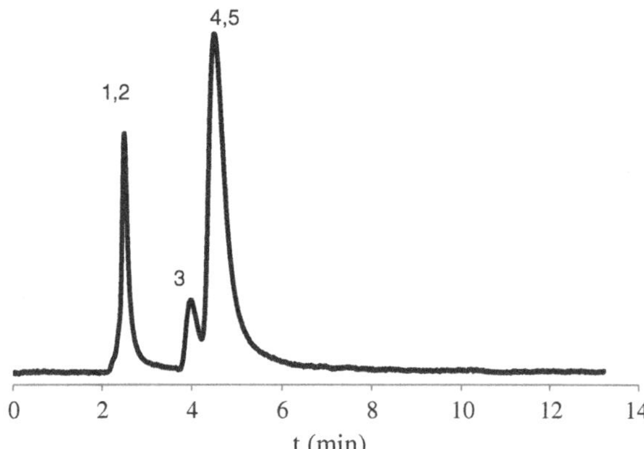

Fig. 3.16 Separation of nucleotides and nucleosides using amorphous carbon as stationary phase. *1* CTP, *2* CMP, *3* CDP, *4* U, *5* C. Mobile phase: 5 % acetonitrile and 50 mM ammonium acetate buffer. Flow rate: 4 μL/min. UV detection at 254 nm

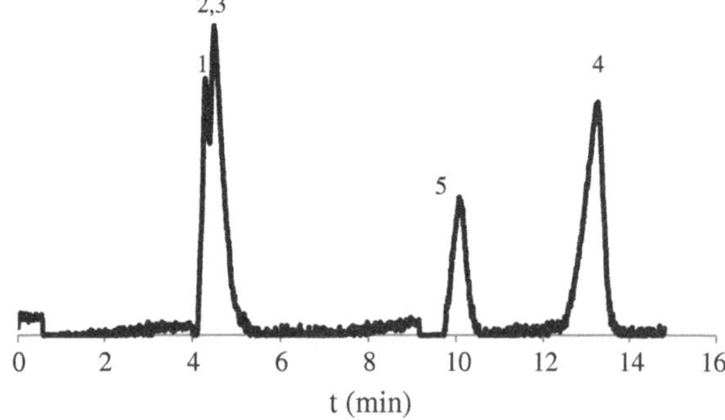

Fig. 3.17 Separation of nucleotides and nucleosides using Hypercarb as stationary phase. *1* CTP, *2* CMP, *3* CDP, *4* U, *5* C. Mobile phase: 20 % acetonitrile and 50 mM ammonium acetate buffer. Flow rate: 4 μL/min. UV detection at 254 nm

3.3.6 Separation of Amino Acids

Amino acids are ionic analytes and often separated using capillary electrophoresis [29], HPLC [30] and TLC [31]. A commonly-used method is to label the amino acids with dyes [32–36] so that the amino acids can be detected using UV or

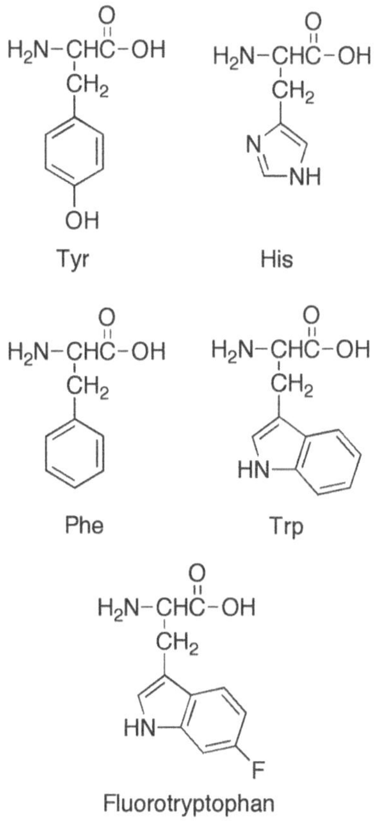

Fig. 3.18 Structures of Tyr, His, Phe, Try and fluorotryptophan

fluorescence detectors [34]. Another advantage of chemical derivatization is that the dye molecules are usually hydrophobic and can improve the retention of the labeled amino acids under reversed-phase conditions. However, the derivatization has disadvantages as well. The derivatization reaction often introduces impurities [35], lowers the stability [36], and requires extra steps before the chromatography separation. The separation of unlabeled amino acids is difficult using reversed-phase chromatography because the stationary phases are hydrophobic and cannot retain the ionic solutes. However, the edge-plane carbon is more polar. Therefore, the separation of amino acids on edge-plane carbon under reversed-phase conditions was studied herein. Trp, Tyr, Phe and His have very similar structures that contain aromatic rings. Fluorotryptophan is an inhibitor of tryptophan hydroxylase [37] and has a similar structure to Trp. The structures of the analytes are shown in Fig. 3.18.

A mixture containing Trp, Tyr, Phe, His and fluorotryptophan were separated using edge-plane carbon phase under reversed-phase conditions. Different types of buffer salts, including acetate, formate and citrate, were used to optimize the selectivity. The pK_a of imidazole side chain of His is about 6.0 [38]. To avoid the separation condition near the pK_a, the pH of the buffer solution was adjusted to 4. The buffer concentration, 50 mM, was chosen because the higher concentration

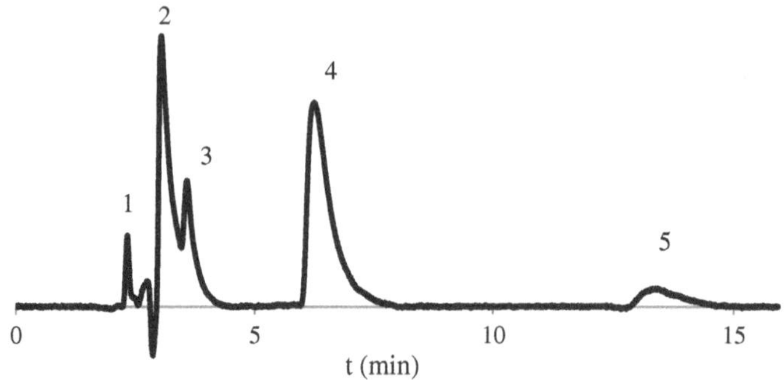

Fig. 3.19 Separation of amino acids using edge-plane carbon stationary phase. 30 % ACN in formate buffer (50 mM, pH = 4). UV detection at $\lambda = 254$ nm. *1* Tyr, *2* His, *3* Phe, *4* Trp, *5* fluorotryptophan

provides higher buffer capacity and better reproducibility. Tyr and Phe are the critical pair and can be separated only when formate buffer was used. Figure 3.19 shows the separation using formate buffer. Tyr and Phe are not fully resolved due to the tailing of the bands.

TFA has been used as an electronic modifier for the separation of ionic solutes using Hypercarb to improve the peak shape previously [39]. TFA is acidic and it can compete with the strong acidic sites on the stationary phase. As a result, the strong interactions between the basic analytes and the stationary phase is decreased and the peak shape is improved [40]. Additionally, TFA may act as a weak ion-pair reagent. The separation of the ionic compound on a reversed-phase column can be improved [40]. Therefore, TFA was added as an organic additive to improve the separation. Figure 3.20 shows the separation with 0.1 % TFA. The peak tailing was greatly improved and Try and Phe were well separated. The plate height of Phe is 14 μm while the minimum plate height obtained in Sect. 3.3.2 is 30 μm. The improvement of the efficiency is due to the addition of TFA.

Separation of amino acids on a Hypercarb column has been reported using ion pair chromatography [41]. Although an ion pair reagent is involved in ion pair chromatography to improve the retention of solutes, the selectivity should not change due to the ion pair reagent. It has been reported that changing the ion pair reagent does not change the selectivity in ion pair chromatography [42]. This result shows that the selectivity is controlled by the type of analytes. Although the separation on Hypercarb was performed using ion pair chromatography we are still able to compare the selectivity of the separation with a reversed-phase condition. The elution order of Tyr and Phe is different from that using the edge-plane carbon phase. Because acetonitrile was used for both Hypercarb and the edge-plane carbon phase as an organic modifier the reason for the difference in elution order is that the selectivity of the edge-plane carbon is different from Hypercarb. We also separated these amino acids on Hypercarb column using the buffers that were used on the edge-plane carbon phase. However, Tyr and Phe co-eluted using these common buffers. A representative chromatogram is shown in Fig. 3.21.

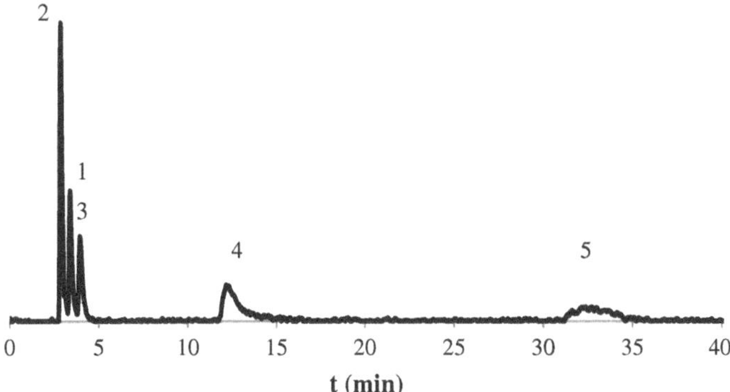

Fig. 3.20 Separation of amino acids using edge-plane carbon stationary phase. 10 % ACN in formate buffer (50 mM, pH = 4), 0.1 % TFA. UV detection at λ = 254 nm. *1* Tyr, *2* His, *3* Phe, *4* Trp, *5* fluorotryptophan

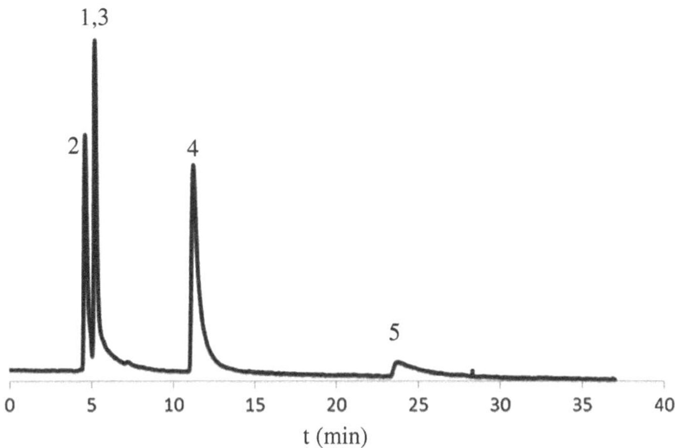

Fig. 3.21 Separation of amino acids using Hypercarb stationary phase. 30 % ACN in formate buffer (50 mM, pH = 4), 0.1 % TFA. UV detection at λ = 254 nm. *1* Tyr, *2* His, *3* Phe, *4* Trp, *5*-fluorotryptophan

3.3.7 Column to Column Reproducibility

The column to column reproducibility is critical for scale up and commercialization of the edge-plane carbon phase. Therefore, three batches of the edge-plane carbon coated silica gel particles prepared from three individual fluidized coating and furnace heating processes were packed into three micro columns with the same length. The retention times of the nucleotides and nucleosides using three columns

Table 3.5 Reproducibility of three batches of edge-plane carbon

	CMP	CDP	U	C
#1	3.74	6.73	9.71	10.4
#2	3.77	6.77	9.59	10.51
#3	3.66	6.83	10.65	11.59
Average	3.72	6.78	9.98	10.83
SD	0.06	0.05	0.58	0.66
RSD	0.02	0.01	0.06	0.06

Mobile phase 5 % acetonitrile and 50 mM ammonium acetate buffer, flow rate: 6 μm/min, 254 nm UV detection

were shown in Table 3.5. The relative standard deviation of the three batches is 1–6 %. Our result is comparable to the reported carbon coated stationary phase using CVD method [8].

3.4 Conclusion

The homogeneous edge-plane carbon stationary phase was prepared and fully characterized. The high-resolution TEM and XPS results confirmed the formation of the edge-plane carbon on the silica gel support. The amount of the carbon load was optimized to achieve the highest carbon coverage while avoid the agglomeration of the particles. LSER study showed that the edge-plane carbon surface is more polar and the interactions on edge-plane carbon are significantly different than the interactions on Hypercarb. The unique selectivity was further demonstrated by the separation of a mixture of nucleosides. The best selectivity was achieved by using the edge-plane carbon phase compared to amorphous carbon and Hypercarb. By adding TFA the peak shape was greatly improved for the separation of amino acids using the edge-plane carbon phase under reversed-phase conditions. This study showed a great potential of the edge-plane carbon as a stationary phase that provides unique selectivity especially for more polar analytes.

References

1. Cserhati T (2009) Biomed Chromatogr 23:111
2. Di Corcia A, Samperi R, Marcomini A, Stelluto S (1993) Anal Chem 65:907
3. Knox JH, Ross P (1997) Adv Chromatogr 37:73
4. Unger KK (1983) Anal Chem 55:361A
5. Jackson PT, Kim T, Carr PW (1997) Anal Chem 69:5011
6. Knox JH, Kaur B, Millward GR (1986) J Chromatogr 352:3
7. Weber TP, Carr PW (1990) Anal Chem 62:2620

8. Paek C, McCormick AV, Carr PW (2010) J Chromatogr A 1217:6475
9. Engel TM, Olesik SV, Callstrom MR, Diener M (1993) Anal Chem 65:3691
10. Jenkins GM, Kawamura K (1971) Nature 231:175
11. Pereira LJ (2008) Liq Chromatogr R T 31:1687
12. Jian K, Shim H-S, Schwartzman A, Crawford GP, Hurt RH (2003) Adv Mater 15:164
13. Jian K, Shim H-S, Tuhus-Dubrow D, Bernstein S, Woodward C, Pfeffer M, Steingard D, Gournay T, Sachsmann S, Crawford GP, Hurt RH (2003) Carbon 41:2073
14. Forgacs EJ (2002) J Chromatogr A 975:229
15. West C, Elfakir C, Lafosse M (2010) J Chromatogr A 1217:3201
16. Wahab MF, Ibrahim MEA, Lucy CA (2013) Anal Chem 85:5684
17. Shearer JW, Ding L, Olesik SV (2007) J Chromatogr A 1141:73
18. Jian K, Xianyu H, Eakin J, Gao Y, Crawford GP, Hurt RH (2005) Carbon 43:407
19. Leon Y, Leon CA, Solar JM, Calemma V, Radovic LR (1992) Carbon 30:979
20. Paek C, Huang Y, Filgueira MR, McCormick AV, Carr PW (2012) J Chromatogr A 1299:129
21. Vitha M, Carr PW (2006) J Chromatogr A 1126:143
22. Trone MD, Khaledi MG (2000) J Chromatogr A 886:245
23. Rittenhouse CT, Olesik SV (1996) J Liq Chromatogr R T 19:2997
24. Xing J, Apedo A, Tymiak A, Zhao N (2004) Rapid Commun Mass Spectrom 18:1599
25. Yang FQ, Li DQ, Feng K, Hu DJ, Li SP (2010) J Chromatogr A 1217:5501
26. Philibert GS, Olesik SV (2011) J Chromatogr A 1218:8222
27. Zhou T, Lucy CA (2008) J Chromatogr A 1187:87
28. Feng H, Wong N, Wee S, Lee M (2008) J Chromatogr B 870:131
29. Desiderio C, Lavarone F, Rossetti DV, Messana I, Castagnola M (2010) J Sep Sci 33:2385
30. Molnar-Perl I (2003) J Chromatogr A 987:291
31. Lu T, Olesik SV (2013) J Chromatogr B 912:98
32. Cheng Y, Dovichi NJ (1988) Science 242:562
33. Cheng Y, Dovichi NJ (2010) J Sep Sci 33:2385
34. Haynes PA, Sheumack D, Kibby J, Redmond JW (1991) J Chromatogr 540:177
35. Cohen SA (1990) J Chromatogr 512:283
36. DeJong C, Hughes CJ, Wieringen EV, Wilson KJ (1982) J Chromatogr 241:345
37. Nicholson AN, Wright CM (1981) Neuropharmacology 20:335
38. Liu T, Ryan M, Dahlquist FW, Griffith OH (1997) Protein Sci 6:1937
39. Elfakir C, Dreux M (1996) J Chromatogr A 727:71
40. Guo D, Mant CT, Hodges RS (1987) J Chromatogr 386:205
41. Petritis K, Chaimbault P, Elfakir C, Dreux M (2000) J Chromatogr A 896:253
42. Inchauspe G, Delrieu P, Dupin P, Laurent M, Samain D (1987) J Chromatogr 404:53

Chapter 4
Preparation of Homogeneous Basal-Plane Carbon Nanorods and Its Application for Solid Phase Extraction

Abstract A method for the preparation of basal-plane carbon nanorods was developed. A preliminary study of solid-phase extraction selectivity of basal-plane carbon nanorods was performed.

4.1 Introduction

Solid phase extraction (SPE) is an important method used for sample pre-treatment and purification [1]. SPE uses a sorbent to selectively absorb the analytes from a complex sample solution, and then the absorbed analytes may be desorbed in a strong solvent or at a high temperature for further analysis. Selective enrichment of the interested analytes improves the sensitivity of the following tests. The sorbents can absorb specific impurities and remove them from the solution as a purification method as well. Carbon as a sorbent for SPE shows good selectivity toward aromatic analytes due to the dispersive interaction and the dipole-induced dipole interaction [2]. Commonly-used carbon sorbents in SPE include carbon black [2], activated carbon [2], and Hypercarb [3]. They have been applied to water analysis [2], pesticide [4], organic pollutants [5], and food analysis [6], etc. Nanostructured carbon materials, such as carbon nanotubes [7], have also been used as solid sorbents due to their high surface area. The advantage of high surface area is that it can improve the extraction efficiency [7].

Currently-used carbon sorbents are amorphous carbon containing both edge-plane sites and basal-plane sites, similar to the carbon stationary phases for liquid chromatography that were discussed in Chap. 3. For currently used preparation methods, the composition of edge-plane sites and basal-plane sites is not controlled.

Similar to the carbon LC stationary phases discussed in Chap. 3, the alignments of graphite layers are expected to affect the type of interactions between the carbon sorbents and the analysts. The edge-plane sites are chemically more reactive and may contain different types of functionalities [8]. In addition, the edge-plane surface is highly hydrophilic [9]. The interactions with polar analytes are favored on the edge-plane

T. Lu, *Nanomaterials for Liquid Chromatography and Laser Desorption/Ionization Mass Spectrometry*, Springer Theses,
DOI: 10.1007/978-3-319-07749-9_4,

surface. On the other hand, basal-plane sites are more stable and hydrophobic [10]. Therefore, by controlling the alignment of graphite layers, the surface property of carbon can be tuned, and the selectivity can vary when used as sorbents in SPE.

The selectivity of the aromatic solutes has been studied by using homogenous edge-plane carbon sorbents [11]. The edge-plane carbon shows higher extraction efficiency to the more polar analytes, phenol and p-cresol; for the hydrophobic analytes, benzene and ethylbenzene, edge-plane carbon shows lower extraction efficiency compared to amorphous carbon. This result further demonstrates that the edge-plane carbon surface is more polar and it provides different SPE selectivity compared to the amorphous carbon nanorods.

The synthesis of the edge-plane carbon nanorods has been reported using a template method [12]. The advantage of using nanorods is their large surface area, which provides higher extraction efficiency. An anodic aluminum oxide (AAO) membrane with well defined pore structure was used as the template. The carbon precursor (AR meso phase) solution was added onto the membrane and the solution infused into the pores by capillary action. After pyrolysis the carbon precursor turned to carbon. Then the template was dissolved and the carbon nanorods were obtained [12]. AR meso phase can form ordered alignment at its liquid crystalline state. Depending on the type of substrates, edge-on or face-on alignment can be formed [12]. The formation of the face-on anchoring state of AR meso phase has been observed on the substrates including highly ordered pyrolytic graphite (HOPG), silver and platinum [13]. The preparation of the carbon/carbon composite nanofibers using HOPG as a substrate has been reported [14]. The basal-plane carbon and HOPG formed a core-shell structure. However, the HOPG shell cannot be removed. The carbon nanorods with basal-plane surface have never been reported. For a fair comparison, the surface area of the edge-plane carbon and the basal-plane carbon should be similar. By using the same template, a similar surface area should be expected.

The AAO membrane needs to be modified before it can be used as the template for the basal-plane carbon synthesis. Direct use of the AAO membrane yields edge-plane carbon nanorods. Silver is a substrate that provides a face-on anchoring state [13] and silver plating technique has been well developed [15]. Therefore, we coated the AAO membrane with silver before it was used as a template for carbon preparation. Silver plating on alumina surface can be achieved by reducing silver ions to silver in solution phase [15]. There are several ways to achieve this goal. First, an electrolysis method can be used. Briefly, a reduction potential can be applied to the solution to reduce silver ion to silver metal [16]. This method, however, requires extra apparatus. Secondly, a non-electrolysis deposition method has also been reported [17]. This method uses a weak reducing reagent, glucose, to reduce silver complex, such as $Ag(NH_3)_2^+$, to silver slowly. Silver deposits on the AAO membrane due to the interaction with the active sites [17].

In this work, basal-plane carbon nanorods with similar dimension to the edge-plane carbon nanorods were prepared. A new preparation method using the silver coated AAO template was developed. Aromatic analytes were used to evaluate the basal-plane carbon nanorods as a sorbent for SPE. The extraction efficiency was evaluated and compared with the edge-plane and amorphous carbon. Based on the preliminary results we demonstrate that the selectivity of SPE can be improved by using the homogenous ordered carbon materials as sorbents.

4.2 Experiments

4.2.1 Reagents

AR meso phase pitch was purchased from Mitsubishi Gas Chemical America (New York, NY). Whatman AAO membranes with pore sizes of 200 nm and thickness of 60 μm were purchased from VWR (Batavia, IL). Silver nitrate, L-tartaric acid, glucose, phenol, p-cresol, benzene, ethylbenzene, and pyridine were purchased from Sigma Aldrich (St. Louis, MO). Forming gas (5 % hydrogen and 95 % nitrogen, standard grade) was purchased from Praxair. Methanol (HPLC grade) was purchased from Macron Chemicals (St. Louis, MO). Ammonium hydroxide, nitric acid and sodium hydroxide were purchased from Fisher Scientific. Water was in-house 18 MΩ nanopure.

4.2.2 Preparation of the Basal-Plane Carbon Nanorods

The silver plating on the AAO membrane was performed following a reported procedure with modifications [17]. The AAO membrane was first cleaned thoroughly with hexane, chloroform, and acetone, and dried under vacuum for more than 8 h in order to remove the gaseous impurities in the pores. The silver complex solution was prepared by the following procedure: one drop of 6.0 M NaOH was added to 100 mL $AgNO_3$ solution (4.25×10^{-3} M), and then ammonium hydroxide was added to completely dissolve the brown precipitant. The silver complex solution was then filtered. The pH of the silver complex solution was adjusted to 9 using ammonium hydroxide. A glucose solution (2.27×10^{-2} M) with tartaric acid (2.67×10^{-3} M) was added to the silver complex solution. Then the cleaned AAO membrane was immersed into the solution. The mixture was kept at 50 °C for 2 days until the AAO membrane turned black or dark brown.

AR MP precursor solution (0.5 % (w/w), in pyridine) was added onto the modified AAO membrane. Carbon formed after heating under a forming gas flow in a tube furnace (Lindberg/Blue M Asheville NC, Model: STF55346C-1). The temperature program was following a previously optimized procedure [11]. In brief, the temperature was increased to 300 °C (30 °C/min), and then the temperature was held at 300 °C for 4 h. Then the temperature was increased to 700 °C (3.4 °C/min) and was held for 1 h. After heating, the furnace was slowly cooled down to the room temperature. The membrane was first dissolved in a 6 M NaOH solution to remove aluminum oxide, and washed with water until neutral. Then Ag/carbon nanorods were dissolved in 25 % HNO_3 to remove the silver followed by water wash until the solution was neutral. The carbon nanorods were dried after washing.

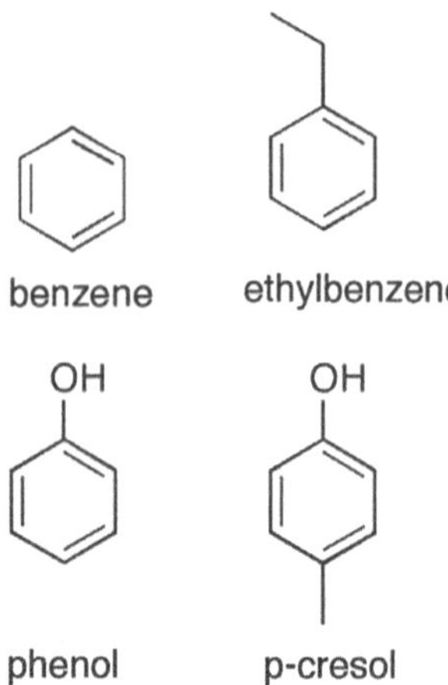

Fig. 4.1 The structures of the SPE analytes

4.2.3 Solid Phase Extraction with the Basal-Plane Carbon Nanorods

Benzene, elthylbenzene, phenol, and p-cresol were used as analytes. Their structures are shown in Fig. 4.1. These analytes are all common pollutants in the environmental and industrial byproducts [18]. Approximately 1–5 mg of the basal-plane carbon was weighed accurately and transferred to a 2 mL glass HPLC sample vial. Then 1.5 mL of the analyte solution (10 μg/mL) was added to the vial. The vial was then capped and the suspension was agitated for 1 h using a Dremel engraver (model 290-01) as a vortex device. The vial was then removed from the Dremel engraver, and the solution was separated from the carbon sorbents by centrifugation for 3 min. The concentration of the solution was determined via HPLC. The LC system was consisted of a Kromasil C-18 (4.6 × 150 mm, 5 μm packing) HPLC column (Sigma Aldrich, St. Louis, MO), a Shimadzu SPD 20A UV-Vis detector (Columbia, MD) and an ISCO 260D syringe pump (Lincoln, NE). 65:35 methanol:water was used as the mobile phase at a flow rate of 1.00 mL/min.

4.3 Results and Discussion

4.3.1 Preparation and Characterization of the Basal-Plane Carbon Nanorods

In order to prepare the basal-plane carbon nanorods, we used a template method. The AAO templates were coated with silver on the surface. Surfaces, such as silver and platinum, favor the formation of the basal-plane anchoring state of the AR meso phase precursor [13]. It has been proposed that the hydrophobicity and the flatness of the surface of silver or platinum ensure the formation of the face-on anchoring state [13]. Silver plating was achieved by following a reported method using a mild reducing agent (glucose) and the silver ammonium complexes [17]. This method is simple and environmental friendly with the nontoxic chemicals. This reaction is similar to the Tullens' test of aldehydes. The aldehyde group of glucose can react with the silver complex and form silver metal. The reaction rate is controlled by the pH of the solution [17]. Higher pH can increase the reaction rate. Fast reaction will result in precipitation of silver in the solution. To avoid this, the solution pH was adjusted to 9. When the reaction is slow the silver reduction can only occur on the active sites of AAO templates. If the reaction was performed at room temperature it took about 2 weeks to form silver coating. At 50 °C the silver coating can be formed in 2 days. The container of the reaction solution is important for the silver deposition. First, the container should be made of plastic since silver coating can form on the glassware as well. Secondly, the container must be clean. Certain sorts of impurities dissolving in the reaction solution can lead to silver precipitation. The formation of the silver coating is indicated by the change of color from white to black or dark brown.

The silver coated AAO membranes were subsequently used as the templates to prepare the basal-plane carbon nanorods. For the preparation of the edge-plane carbon nanorods, 1 % AR meso phase in pyridine was used [11]. However, when 1 % AR meso phase was used for the preparation of the basal-plane carbon nanorods very few nanorods were observed under TEM. The reason may be after the silver coating, the pore size of AAO membranes is expected to decrease. It may be difficult for the AR meso phase solution to diffuse into the pores. To prepare the basal-plane carbon nanorods, the concentration of AR meso phase was decreased to 0.5 %.

The high resolution TEM was used to visualize the carbon alignment on the surface of the carbon nanorods. Figure 4.2 shows the TEM image of the basal-plane carbon nanorods. It is clear that the surface graphene layer alignment was face on.

4.3.2 Solid Phase Extraction

The basal-plane carbon nanorods were used as sorbents for solid phase extraction. Because the carbon nanorods have small diameter (200 nm) the amount of the carbon sorbents used was small (1–5 mg/1.5 mL solution). The extraction efficiency is defined as the ratio of the amount of analytes adsorbed on the sorbent and the mass of the sorbent. The amount of the analytes adsorbed was determined by the concentration of the solution after the extraction using LC.

Benzene, ethylbenzene, phenol and p-cresol in aqueous solutions were used as analytes. Figures 4.3, 4.4, 4.5 and 4.6 shows the extraction efficiency of all analytes by using three different sorbents including the basal-plane, the edge-plane

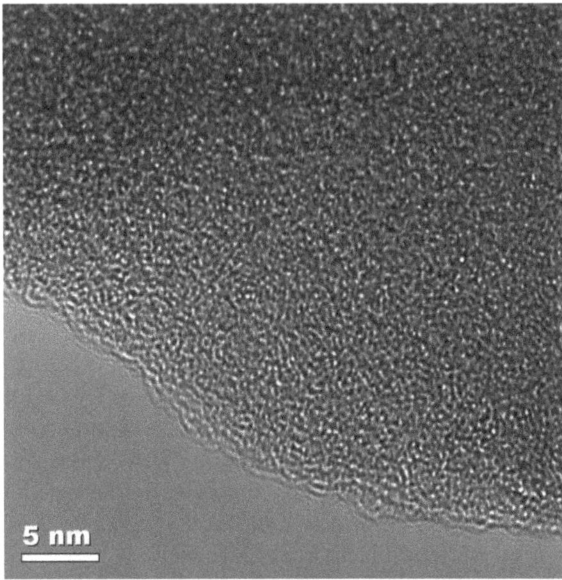

Fig. 4.2 High resolution TEM image of the basal-plane carbon nanorods

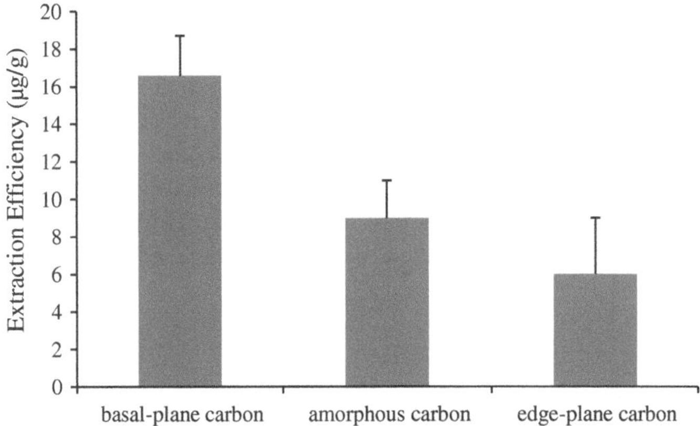

Fig. 4.3 Extraction efficiency of benzene

and the amorphous carbon nanorods. First, we found that the extraction efficiency of benzene and ethlybenzene is higher on the basal-plane carbon than p-cresol and phenol. The interactions between the analytes and the basal-plane of carbon surface are largely through the π-π interactions. Because of the large amount of exposed basal-plane carbon surface, the π-π interactions are stronger. Phenol and p-cresol, however, contain polar functional group, hydroxyl group, through which the hydrogen bond with water molecules can form. The strong interactions

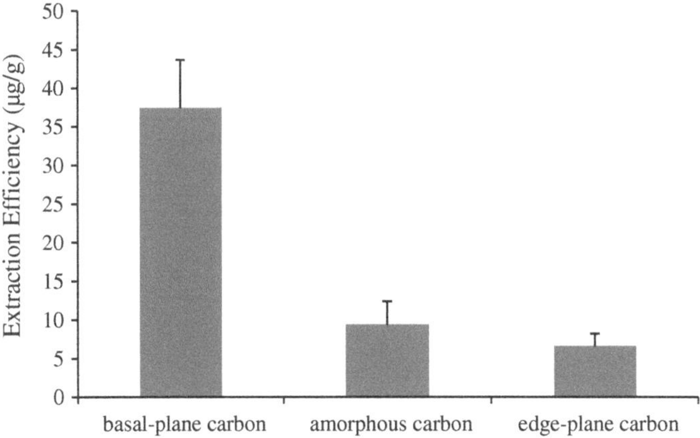

Fig. 4.4 Extraction efficiency of ethylbenzene

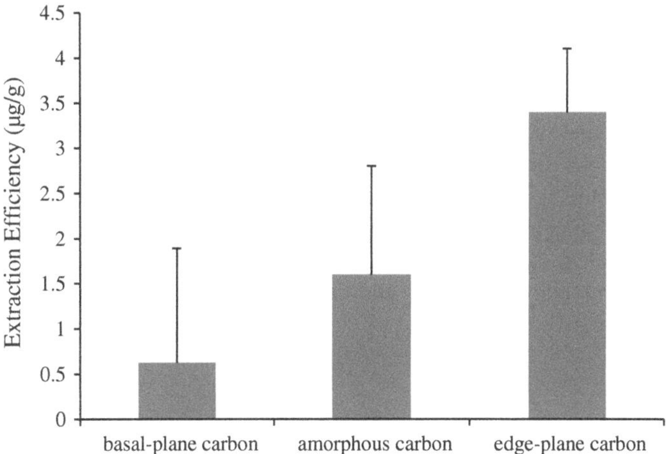

Fig. 4.5 Extraction efficiency of phenol

between hydroxyl groups and water molecules decrease the extraction efficiency. Secondly, we found that compared to basal-plane carbon when the edge-plane carbon was used as a sorbent the extraction efficiency of benzene and ethlybenzene was much smaller [11]. The surface of the edge-plane carbon is composed of the edge of the graphene layers. No aromatic carbon rings are present for the π-π interactions with the analytes, which decreases the extraction efficiency to hydrophobic solutes. Thirdly, the extraction efficiency of the polar analytes such as phenol and p-cresol is higher on the edge-plane carbon than that on the basal-plane carbon. The edge-plane carbon is more polar, which can have the dipole interactions with polar functional groups in the analytes. In addition, there may be

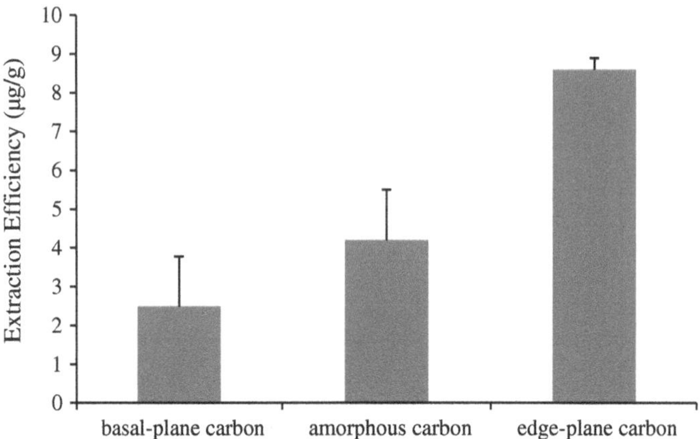

Fig. 4.6 Extraction efficiency of p-cresol

organic functional groups, such as hydroxyl and carboxylic groups, on the edge-plane carbon [19]. These functional groups can have hydrogen bond interactions with the hydroxyl groups of the analytes as well. Therefore, the extraction efficiency of the polar analytes, phenol and p-cresol, is higher on the edge-plane carbon than on the basal-plane carbon (Figs. 4.5 and 4.6). The results were also compared with the amorphous carbon nanorods that contain both the edge-plane and the basal-plane sites. The extraction efficiency of the amorphous carbon is between that of the basal-plane carbon and the edge-plane carbon for all analytes. Due to the different extraction selectivity of the edge-plane carbon and the basal-plane carbon, by simply changing the carbon alignment the extraction efficiency can also be changed.

4.4 Conclusion

In this chapter, we have prepared the basal-plane carbon nanorods using the silver coated AAO membranes as templates. High resolution TEM confirms the formation of the basal-plane carbon surface. The carbon nanorods were used as sorbents for solid phase extraction of the aromatic analytes in aqueous solutions. The SPE results were compared with the edge-plane and amorphous carbon nanorods with similar dimensions. The preliminary results reveal that due to the difference in carbon alignment, the basal-plane carbon shows higher extraction efficiency than the edge-plane carbon for hydrophobic analytes; the edge-plane carbon shows higher extraction efficiency for polar analytes. The extraction efficiency of the amorphous carbon is between the two carbon phases. Our results indicate that the selectivity of SPE can be improved by using the homogenous ordered carbon as sorbents for different types of analytes. It is important to note that our results provide an important evidence on the interactions between the different types of carbon and analytes, which are very important for almost all separation techniques using carbon as the stationary phase, e.g. SPE, LC, GC, CE, etc.

References

1. Buszewski B, Szultka M (2012) Crit Rev Anal Chem 42:198
2. Hennion M (2000) J Chromatogr A 885:73
3. Michel M, Buszewski B (2009) Adsorption 15:193
4. Pyrzynska K (2011) Chemosphere 83:1407
5. Niu H, Cai Y (2012) Advance in nanotechnology and the environment. Pan Stanford Publishing Pte Ltd, Singapore, pp 79–136
6. Li G, Ma G (2011) Sepu 29:606
7. Herrera AV, Gonzalez-Curbelo MA, Hernandes-Borges J, Rodriguez-Delgado MA (2012) Anal Chim Acta 734:1
8. Phillips R, Vastola FJ, Walker PL Jr (1970) Carbon 8:197
9. Ye H, Naguib N, Gogotsi Y, Yazicioglu AG, Megaridis CM (2004) Nanotechnology 15:232
10. Darnstadt H, Ry CH (2003) Carbon 41:2662
11. Zewe J (2012) The development of novel nanomaterials for separation science. Dissertation
12. Jian K, Shim H, Schwartzman A, Crawford GP, Hurt RH (2003) Adv Mater 15:164
13. Jian K, Shim H, Tuhus-Dubrow D, Bernstein S, Woodward C, Pfeffer M, Steingart D, Gournay T, Sachsmann S, Crawford GP, Hurt RH (2003) Carbon 41:2073
14. Chan C, Crawford G, Gao Y, Hurt R, Jian K, Li H, Sheldon B, Sousa M, Yang N (2005) Carbon 43:2431
15. Blair A (2011) Met Finish 109:239
16. Schlesinger M (2010) Modern electroplating, 5th edn. Wiley, New York, pp 131–138
17. Zhang S, Xie Z, Jiang Z, Xu X, Xiang J, Huang R, Zheng L (2004) Chem Commun 1106
18. Shirey RE (2000) J Chromatogr Sci 38:279
19. Pereira L (2008) J Liq Chromatogr R T 31:1687

Chapter 5
Electrospun Nanofibers as Substrates for Surface-Assisted Laser Desorption/Ionization and Matrix-Enhanced Surface-Assisted Laser Desorption/Ionization Mass Spectrometry

Abstract Nanofibrous substrates prepared by electrospinning were studied for improvement of the performance of laser desorption/ionization mass spectrometry. Polymers, polyacrylonitrile, poly(vinyl alcohol), and SU-8 photoresist, and carbon substrates were studied. The results show that poly(ethylene glycol) with a molecular weight as high as 900,000 Da was detected when the carbon nanofibrous substrate was used without other matrix (SALDI). Low detection limit of 400 amol was achieved for angiotensin I when the carbon substrate was used with conventional matrix (ME-SALDI).

5.1 Introduction

Matrix-assisted laser desorption/ionization (MALDI) is a widely used soft ionization method for nonvolatile and thermally labile large molecules, such as biomolecules and synthetic polymers [1, 2]. MALDI uses organic matrix molecules that absorb and transfer laser energy to the analytes which are co-crystallized with the matrix molecules on a substrate surface. Analyte molecules are then desorbed, ionized and detected using mass spectrometry (MS). However, the use of organic matrices causes an inhomogeneous distribution of the analyte crystals which leads to the formation of "sweet spots" where the signal to noise ratio (S/N) is the highest. Searching for the "sweet spot" can become time consuming and automation of the technique can be difficult. Because the quality of the results is largely dependent on the preparation step involving the use of matrices in the MALDI analysis, the methods used to prepare homogenous and reproducible samples for MALDI have been studied and developed [3–7].

Surface-assisted laser desorption/ionization (SALDI) was first reported by Sunner et al. [8] as a matrix-free laser desorption/ionization method. Inorganic materials, that are UV absorptive materials such as micro/nano scaled carbon [9–11], silicon [12, 13], and metals [14–16], have been used to generate high quality MS spectra without interference of the signal from the matrix molecules in the low molecular weight range of the spectrum and without "sweet spot" formation.

T. Lu, *Nanomaterials for Liquid Chromatography and Laser Desorption/Ionization Mass Spectrometry*, Springer Theses,
DOI: 10.1007/978-3-319-07749-9_5,

In addition, signal enhancement has been shown to be facilitated by the use of nanostructured materials as the substrate, i.e. the type and nanoscale structure of these materials can greatly impact the quality of the SALDI mass spectra [17–19]. For example, TiN as nano-sized particles has shown SALDI activity while bulk TiN provides no signal [20]. Liquids such as glycerol are usually required for graphite and metal nanoparticles to yield intense signals with high resolution due to the fast heat dissipation in liquid phase [9]. The fast heat dissipation removes the excess heat that can cause lower resolution and more fragmentations. However, the use of liquids often increases the pressure in the ion source, which results in decreased ionization efficiency and also produces background signals [9]. Graphite used in ambient SALDI analyses was recently reported without addition of a viscous liquid; however, the assistance of a direct analysis in real time (DART) ionization source was required [21]. Additionally, typically when nanoparticles are used as the substrate, especially with graphite or metal nanoparticles, they may fly off the target plate and contaminate the instrument; this may lead to discharges and electrical short circuits in the instrument due to the conductivity of the material [9, 10]. Also SALDI mass spectral signals obtained using graphitic substrates are highly sensitive the mixing ratio of the graphite and glycerol [9].

To assist the desorption process, polymer-assisted laser desorption/ionization (PALDI) was introduced. PALDI uses a substrate of synthetic polymers specifically designed and synthesized to provide absorption of ultraviolet radiation [22–24]. As discussed in Chap. 1, matrixes used in PALDI are not commercially available and are only used for analytes with low molecular weight (<1,500 Da) [22]. The detection limit of a peptide using PALDI has been reported to be better than 500 amol [24]. Nevertheless, low cost synthetic polymeric media for SALDI are not currently commercially available [25].

Although a variety of analytes have been studied using matrix-free laser desorption ionization including small molecules [10] and peptides [19]. Only a few reported the application of SALDI for analyses of synthetic polymers [9, 26–29]. The application of SALDI MS has been limited by the inability to ionization high molecular weight synthetic polymers [30] and proteins [31, 32]. The analyte with the highest molecular weight that has been analyzed by SALDI is immunoglobulin G (IgG) with molecular weight of 150,000 Da.

Matrix-enhanced SALDI (ME-SALDI) has been used to increase the ionization efficiency of SALDI by taking advantage of both the proton-donating ability of organic MALDI matrix combined with the efficient photon absorption of porous surface [33]. CHCA and its ionic salt and porous silica substrate have been used for ME-SALDI [34]. The use of ionic matrix improves the signal homogeneity because the ionic liquid matrix does not form the inhomogeneous crystalline spots. ME-SALDI has shown an improvement of detection compared to MALDI. The detection limit of limit of 1,2-dimyristoyl-*sn*-glycero-3-phosphocholine is 10 fmol under ME-SALDI condition [33]. Therefore ME-SALDI has shown great advantage in the analysis and imaging of small molecules. However, other substrates have not been studied thus far.

As mentioned above, the microstructure is crucial for optimal S/N in SALDI. Materials with nanostructures show greatly enhanced S/N compared to the bulk material. Electrospinning is used here to prepare polymer and carbon nanofibers.

This work describes the first use of electrospun polymeric nanofibers, and carbon nanofibers as substrates for SALDI and ME-SALDI study. Polymeric nanofibers with different functionalities were prepared from commercially-available, low-cost polymers, polyacrylonitrile (PAN), polyvinyl alcohol (PVA), and an epoxide-based photoresist (SU-8), and carbon nanofibers were produced through the pyrolysis of SU-8 nanofibers to final temperatures of 600, 800 and 900 °C. For an initial evaluation, a range of analytes were considered including polyethylene glycol (PEG), polystyrene (PS), glucose, arginine, crystal violet, and angiotensin I.

5.2 Experimental

5.2.1 Materials

SU-82100, an epoxide-based negative photoresist, was purchased from MicroChem Corporation (Newton, MA). PVA (99 % + hydrolyzed, M_w 89,000–98,000), PAN (M_w 150,000) and $NaHCO_3$ were purchased from Sigma-Aldrich (St. Louis, MO). SU-8 was diluted to 75 (v/v) % with cyclopentanone (Sigma Aldrich, St. Louis, MO). PVA was dissolved in deionized water (8 %, w/w). Glutaraldehyde (GA, 25 % in water, Baker) and HCl (0.1 M) were added to the aqueous PVA solution prior to electrospinning as a crosslinker and catalyst, respectively [35]. PAN was dissolved in N, N-dimethylformamide (DMF, Sigma Aldrich, St. Louis, MO) with a concentration of 10 % (wt%). Angiotensin I, D-(+)-glucose, L-arginine monohydrochloride, crystal violet, polystyrene 4000 (PS 4000, standard, M_w 4,130), polystyrene 2000 (PS 2000, standard, M_w 2,360), poly(ethylene glycol) (PEG 3400, M_n 3,400), poly(ethylene glycol) (PEG 900000, M_w 900,000), CHCA, 2,5-dihydroxybenzoic acid (DHB), trifluoroacetic acid (TFA), silver trifluoroacetate, and sodium trifluoroacetate were purchased from Sigma-Aldrich (St. Louis, MO). Dithranol (DR) was purchased from Fisher Scientific (Hanover Park, IL). Methanol and tetrahydrofuran (THF) were purchased from Mallinckrodt Chemicals (Phillipsburg, NJ).

5.2.2 Electrospinning

The electrospinning apparatus was previously reported [35] and discussed in Chap. 2. A commercial stainless steel target plate, MSP 96 MicroScout plate (Bruker Daltonics, Ypsilanti, MI) was used as the collector. PAN (10 and 12.5 %, wt%) was electrospun using previously optimized conditions of 20 kV, a distance between the tip and the collector of 20 cm, and a flow rate of 1.5 mL/h with relative humidity less than 50 % [36]. Porous PAN was prepared following a reported procedure [37]. Briefly, 5 or 10 % of $NaHCO_3$ (wt%) was mixed and homogeneously dispersed in 10 % PAN solution. The electrospinning condition was the same as the one used for pure PAN [36]. After electrospinning the fiber mat was peeled off from the collector and immersed into 10 % HCl solution to remove $NaHCO_3$ and create pores. Then the mat was washed with nano pure water thoroughly to remove the salt and HCl. When the mat was still wet it was spread onto the target plate. The fiber mat adhered to the plate when it was dry. The in situ crosslinked PVA was prepared following the procedure described in Chap. 2 [35]. The SU-8 was electrospun in a dark room due to its UV sensitivity. The previously optimized conditions consisted of a voltage of 9 kV, a distance between the tip and the collector of 10 cm, and a flow rate of 1. 2 mL/h [38]. The obtained SU-8 mat was crosslinked under UV light (Sunray 400SM UV flood light, Uvitron International West Springfield, MA) for about 5 min [39]. For all polymers, the electrospinning time varied from 5 to 25 min to obtain a homogeneous coverage of the target plate (dimension app. 53 × 41 mm) and fibrous mats of 2 ± 0.5 mg. The mass of the fibrous mats were determined by the mass change of the target plate before and after electrospinning. For PAN and SU-8, electrospinning about 5 min yielded 2 mg mats, while for crosslinked PVA about 20 min electrospinning time was needed. Because of the mass loss during the pyrolysis, electrospinning times of 15, 25 and 35 min were used to produce 2 mg nanofibrous mats for the carbon nanofibers processed to 600, 800 and 900 °C, respectively.

5.2.3 Fabrication of Carbon Substrates

The carbon nanofibrous substrates were made by pyrolysis of the crosslinked SU-8 precursor. The target plate with electrospun and crosslinked SU-8 was placed in a tube furnace (Lindberg/Blue M Asheville NC, Model: STF55346C-1) and heated under forming gas (5 % H_2 in N_2) at a ramp rate of 1 °C/min for final temperatures of 800 and 900 and 2 °C/min for 600 °C. The final pyrolysis temperature was held for 6 h and cooled to room temperature [39].

5.2.4 Sample Preparation

Polystyrene with silver trifluoroacetate (1:1, w/w) and dithranol were dissolved in THF. Poly(ethylene glycol) with sodium trifluoroacetate (1:1, w/w) and DHB were dissolved in methanol. For the dynamic range study PS 4000 solution (250 mg/mL in THF with 250 mg/mL silver trifluoroacetate) and PS 4000/PS 2000 blend (250 mg/mL in THF with 250 mg/mL silver trifluoroacetate and 5 % PS 2000, mol%) were prepared. Glucose and arginine were dissolved in water (1 mg/mL). Crystal violet was dissolved in methanol (5 mg/mL). Angiotensin I and matrix CHCA were dissolved in acetonitrile/water (1:1, v/v) with 0.1 % trifluoroacetic acid. Dried droplet by the multiple-layer spotting method was used for sample preparation [40]. 0.1 μL of the analyte solution was applied on the target plate for SALDI analysis. Matrix solution was applied over the dried analyte. Due to the limit of the solubility of CHCA and dithranol, multiple layers of the solution were applied to

Table 5.1 Matrix to analyte ratios used in ME-SALDI

Analyte (mg/mL)	Matrix (mg/mL)
PEG 3400 (5)	DHB (40)
PEG 3400 (5)	DHB (80)
PEG 3400 (5)	DHB (160)
PEG 3400 (10)	DHB (40)
PEG 3400 (10)	DHB (80)
PEG 3400 (10)	DHB (160)
PEG 3400 (20)	DHB (40)
PEG 3400 (20)	DHB (80)
PEG 3400 (20)	DHB (160)
PS 4000 (25)	DR (40 × 1)
PS 4000 (25)	DR (40 × 3)
PS 4000 (25)	DR (40 × 5)
PS 4000 (50)	DR (40 × 1)
PS 4000 (50)	DR (40 × 3)
PS 4000 (50)	DR (40 × 5)
PS 4000 (100)	DR (40 × 1)
PS 4000 (100)	DR (40 × 3)
PS 4000 (100)	DR (40 × 5)
Angiotensin I (0.5)	CHCA (10)
Angiotensin I (1)	CHCA (5)
Angiotensin I (1)	CHCA (10)
Angiotensin I (1)	CHCA (10 × 2)
Angiotensin I (2)	CHCA (5)
Angiotensin I (2)	CHCA (10)
Angiotensin I (2)	CHCA (10 × 2)

reach different matrix/analyte ratios (Table 5.1). Matrixes and analytes with different concentrations were prepared in order to study and optimize the matrix to analyte ratio. The electrospun substrates were removed from the target plate after use, and the target plate was reused after thoroughly cleaning the surface with THF, acetone, methanol, and water followed by sonication in methanol for 15 min.

5.2.5 Instrumentation

The mass spectra were collected using a Microflex MALDI-TOF Bruker mass spectrometer in positive ion mode. A nitrogen laser at 337 nm with a pulse width of 3 ns was used. The ion acceleration potential was 20 kV. The laser power used to obtain the spectra and to optimize the S/N ratios was slightly above the ionization threshold. For PEG 900000 linear mode was used. For all the other samples reflector mode was used. Each spectrum was the summation of 120 shots, unless otherwise noted. All of the results were repeated at least three times. FlexAnalysis™ software was used for data analysis. For angiotensin I, Sophisticated numerical annotation procedure (SNAP) was used for the peak pick and S/N determination due to its

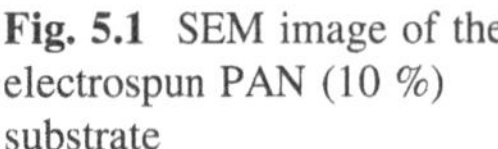

Fig. 5.1 SEM image of the electrospun PAN (10 %) substrate

isotopic pattern, and for all the other analytes CENTROID was used [41]. SNAP is a method for peak picking based on a fitting of multiple isotopic peaks. CENTROID is a method that weighs each data point with respect to its intensity.

A Hitachi S-300 (Hitachi High Technologies, America, Inc., Pleasanton, CA) scanning electron microscope (SEM) was used to obtain images of the nanofibrous substrates. A stainless steel plate of the same dimension as the target plate was used as the collector during electrospinning for the samples used in SEM imaging.

5.3 Results and Discussion

5.3.1 Fabrication of Electrospun Nanofibrous Substrates

Similar random networks of nanofibers were obtained for the polymers and carbon substrates with different diameters ranging from 140 to 460 nm (Fig. 5.1 and Table 5.2). The fiber diameters were determined by the measurements of about 100 fibers. Due to the shrinkage, the diameter of the carbon nanofibers decreases as the final pyrolysis temperature used to produce the carbon increases [42]. For a final temperature of 600 °C a ramp rate of 2 °C/min was used [42]. A mat produced with final processing temperatures of 800 and 900 °C curled away from the stainless steel plate when using this same ramp rate. For this reason a lower ramp rate of 1 °C/min was used and resulted in stronger adhesion of carbon substrate to the plate. For carbon processed to 900 °C bubbles between the substrate and the target plate were still observed.

The target plates with attached electrospun nanofibers were robust due to the strong adhesion of the substrates to the target plate and the high stability of these materials. The circles used to localize the samples on the target plate are still clearly identifiable after electrospinning of polymers (Fig. 5.2) while they disappeared after pyrolysis for carbon substrates (Fig. 5.3). This is highly beneficial for

Table 5.2 Average diameters of the electrospun nanofibers

	Diameter (nm)
Carbon (600 °C)	195 ± 80
Carbon (800 °C)	150 ± 55
Carbon (900 °C)	140 ± 50
PAN (10 %)	460 ± 90
PAN (12.5 %)	880 ± 90
PAN (10 % with 5 % $NaHCO_3$)	470 ± 100
PAN (10 % with 10 % $NaHCO_3$)	430 ± 80
PVA	190 ± 50
SU-8	420 ± 90

Fig. 5.2 Digital photographs of target plates with PAN as substrate

Fig. 5.3 Digital photographs of target plates with carbon nanofibers processed to 600 °C as substrate

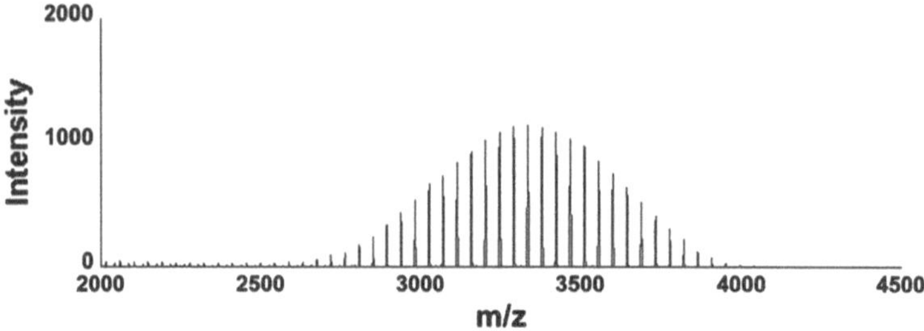

Fig. 5.4 SALDI-TOF mass spectrum of PEG-3400 (20 mg/mL) with the carbon nanofiber substrate processed to 800 °C

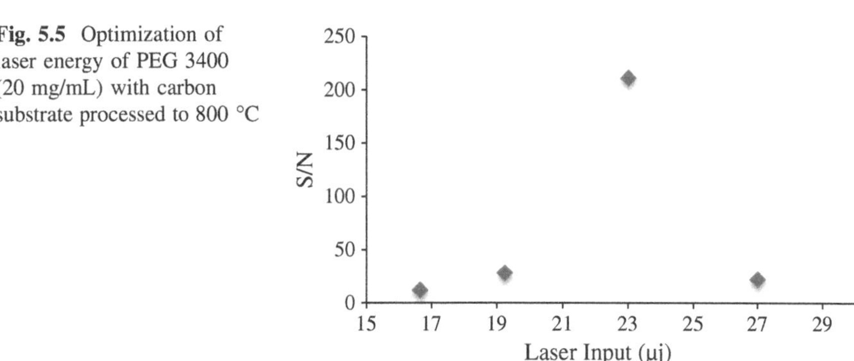

Fig. 5.5 Optimization of laser energy of PEG 3400 (20 mg/mL) with carbon substrate processed to 800 °C

sample preparation. For other SALDI substrates, such as nanostructured gold, an additional patterning step is required to distinguish the sample spotted and unspotted regions [14]. Consequently, sample preparation on the electrospun polymeric target plates was simple, fast, and convenient for SALDI analyses.

5.3.2 SALDI Analysis with Electrospun Nanofibrous Substrates

5.3.2.1 Carbon Substrates

Carbon is a known potential substrate for SALDI applications due to its nearly ideal black body absorption [43]. While desorption and contamination of the instrument has been an issue for graphite and carbon particles [10], it is noteworthy that electrospinning method produces continuous fibers which are much less likely to desorb under laser radiation and contaminate the instrument.

The potential of carbon nanofibers as a SALDI substrate was studied first by using PEG with an $M_w = 3{,}400$ as an analyte (Fig. 5.4). Figure 5.5 shows the optimization of the S/N when different laser energy was used. The optimum laser

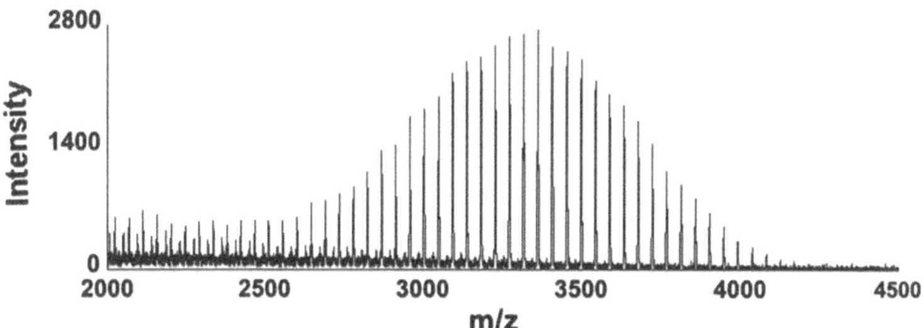

Fig. 5.6 SALDI mass spectrum of PEG 3400 (20 mg/mL) using carbon processed to 800 °C substrate. Laser intensity is 30 μj

intensity for this analysis was 23 μJ using carbon substrate processed to 800 °C. If the laser intensity was further increased fragmentation ion signals and increased noise were observed (Fig. 5.6). When only carbon substrates were used no signals that are similar to the fragment ions were observed. Therefore, these peaks in the lower mass region are from the fragmentation of the PEG sample.

Although signals can be obtained for PEG on the stainless steel target plate directly [44], the use of the spectrum is limited due to the low intensity of the signal. On the nanofibrous carbon substrates, the PEG signal was greatly enhanced compared to that obtained using the stainless steel substrate alone as shown in Figs. 5.4 and 5.7. Previous work has reported using a graphite substrate with glycerol [9] for SALDI analyses of PEG with a similar molecular weight distribution. In comparison, our results showed improved signals without adding a viscous liquid. Because the S/N has not been reported in the literature, we cannot compare the S/N with other SALDI substrates. Carbon nanotubes have been used for SALDI analysis as well. Compared to electrospun carbon nanofibers the diameter of carbon nanotubes is much smaller. However only low molecular weight polymers were analyzed by carbon nanotubes [45]. The laser power used for SALDI using the carbon-based substrates was comparable to the laser power used for MALDI, 23–24 μJ, while the laser power used for LDI using the stainless steel substrate, about 30 μJ, was significantly higher. Polystyrene with similar molecular weight was also tested. No signal was obtained even when the highest laser power was used. Polystyrene may be more difficult to ionize using SALDI compared to PEG. SALDI analysis of PS with only low molecular weight has been reported [27–29].

The signal intensities and S/N obtained from the carbon substrates prepared at different final temperatures were similar (Figs. 5.4, 5.8 and 5.9). However, the spectrum using the carbon substrate processed to 800 °C shows the least amount of fragmentation of the PEG (Fig. 5.4). The optimal laser intensity for the highest S/N was 23 μJ for the carbon substrate processed to 800 °C and was 24 μJ for the carbon substrate processed to 600 and 900 °C. The lower fragmentation may be caused by the slightly lower laser intensity used for carbon processed to 800 °C. Therefore, the fiber diameter is closely related to the ionization softness of PEG. Carbon substrates processed to 800 °C yielded softer ionization than the carbon

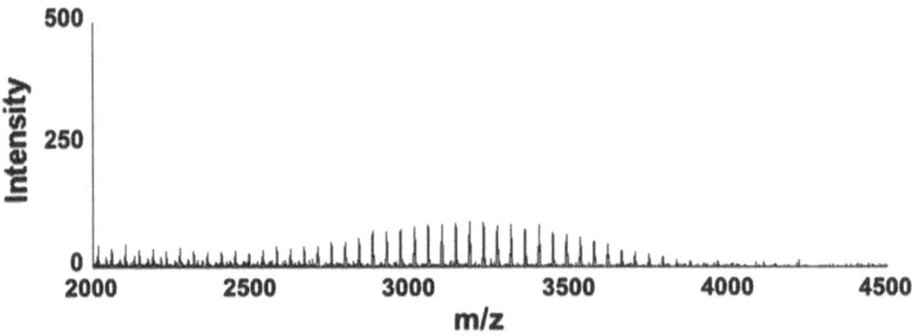

Fig. 5.7 SALDI-TOF mass spectrum of PEG-3400 (20 mg/mL) with stainless steel substrate

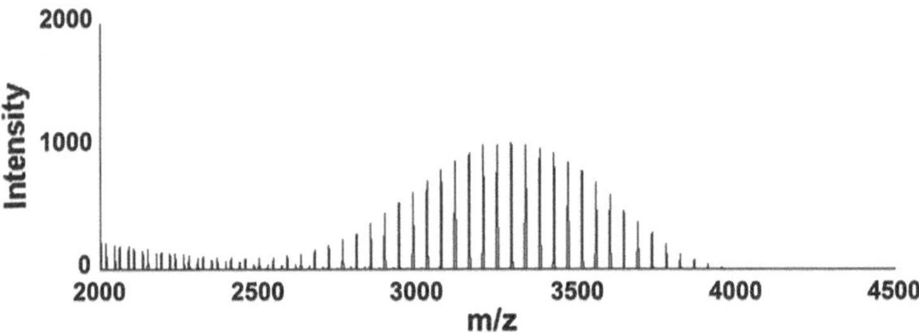

Fig. 5.8 SALDI-TOF mass spectrum of PEG-3400 (20 mg/mL) with the carbon nanofiber substrate processed to 600 °C

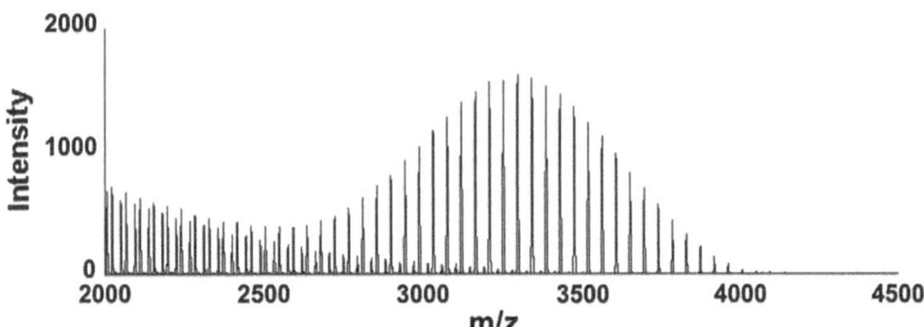

Fig. 5.9 SALDI-TOF mass spectrum of PEG-3400 (20 mg/mL) with the carbon nanofiber substrate processed to 900 °C

substrates processed to 600 and 900 °C, which indicated more efficient energy absorption and transfer by the carbon substrate processed to 800 °C. Even when the laser power of 24 μJ was applied to carbon substrate processed to 800 °C, it still showed a lower amount of fragmentation than the other two carbon substrates. Carbon processed to 900 °C has similar diameters to the carbon nanofibers processed to 800 °C but showed lower S/N. The reason for that may be the poor adhesion of the carbon film to the target plate after pyrolysis at 900 °C. The poor adhesion can cause grounding issue and charge accumulation.

Table 5.3 S/N of SALDI-TOF spectra with electrospun nanofibrous carbon substrates and MALDI for PEG 4300

PEG 4300 (mg/mL)	5	10	20
Carbon (600 °C) (RSD)	4 ± 2 (50 %)	20 ± 9 (45 %)	42 ± 7 (17 %)
Carbon (800 °C) (RSD)	4 ± 2 (50 %)	13 ± 2 (15 %)	35 ± 20 (57 %)
MALDI (RSD)	7 ± 5 (80 %)		

S/N ratios were calculated based on ten spectra with 20 laser shots each

In MALDI signal can be obtained only as laser light is applied at a certain angle relative to the surface of the co-crystals [46]. Therefore, searching for "sweet spot" is often necessary. However, SALDI has been shown to be able to generate more homogenous samples without cocrystallization [10]. Table 5.3 shows the S/N ratios and standard deviations from 10 laser irradiations on randomly chosen spots of one PEG 3400 sample spot using the carbon nanofiber substrates processed to 600 or 800 °C under the same SALDI conditions. Compared to the heterogeneous sample distribution in MALDI, the PEG signal can be observed from all of 10 spots of the sample using carbon substrates without a matrix. The best shot-to-shot reproducibility was obtained from 20 mg/mL PEG for carbon processed to 600 °C while for carbon process to 800 °C the most homogenous sample concentration was 10 mg/mL. The reason was still not clear, but the authors believe that it is related to the different diameters of the two carbon nanofibrous substrates. Relative standard deviations as low as 15 and 17 % were obtained from carbon processed to 800 °C and 600 °C, respectively, while the relative standard deviation was about 80 % for MALDI under optimized experimental condition.

To further investigate the ionization efficiency improvement using carbon nanofibrous substrate, PEG with an $M_w = 900{,}000$ was analyzed using the carbon substrate processed to 800 °C. A previous study reported that low analyte concentrations are required to obtain signal from high molecular weight polymers using MALDI. The cause of this trend was the minimization of the polymer entanglement to produce an ionized polymer [47]. However, no report has been published for SALDI analysis. We found that low concentrations of PEG 900000 were also required. Concentrations above 25 mg/mL did not provide signal. Signals from multiply charged ions were observed with peaks corresponding to ions with two, three, four, and five charges (Fig. 5.10) as expected since multiple charged ions are commonly observed for high mass analytes [32]. Because of the limited mass range of the instrument, molecular ion peaks, m/z = 900,000 with single charge, were not observable. Compared to the spectrum of PEG 3400, the spectrum of PEG 900000 showed lower S/N ratios. This may due to one of two reasons: (1) high mass analytes are harder to ionize and (2) the molar concentration of PEG 900000 (30 μmol) is much lower than the molar concentration of PEG 3400 (4 mmol). To the best of our knowledge, this is the highest mass that has been analyzed by SALDI MS. The mass is much higher than the currently reported upper mass limit of the SALDI technique, 150,000 Da (IgG using HgTe

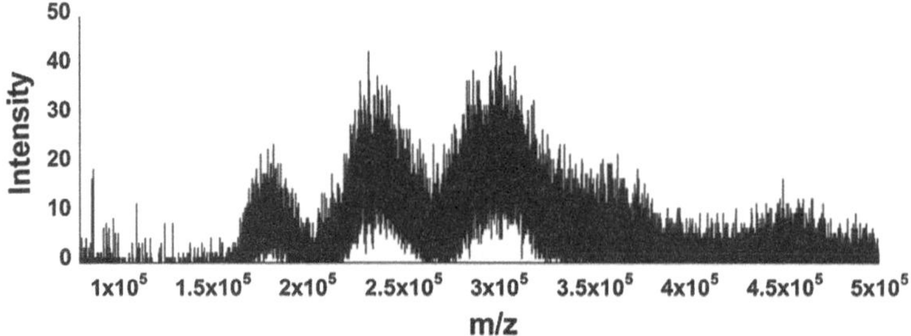

Fig. 5.10 SALDI-TOF mass spectrum of PEG 900,000 (25 mg/mL) with the carbon substrate processed to 800 °C. The spectrum is the summation of 50 laser shots

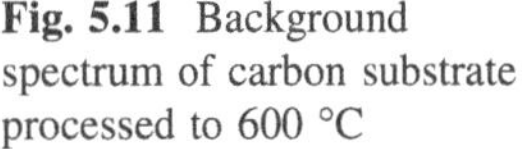
Fig. 5.11 Background spectrum of carbon substrate processed to 600 °C

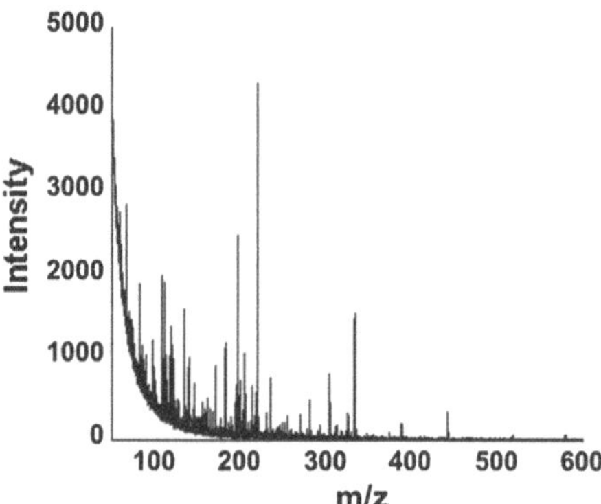

nanostructures) [32]. To the best of our knowledge, this is the highest molecular weight of PEG that has been analyzed by laser desorption MS method [48].

SALDI MS has been reported to be a powerful tool for small molecule analyses. Here we demonstrate that electrospun nanofibrous carbon substrate can also be applied for desorption/ionization of small molecules with negligible background. Nanofibrous carbon substrate processed to 800 °C was used because the background signals generated by carbon processed to 800 °C was lower than 600 °C processed carbon. Figure 5.11 shows the background spectra of carbon processed to 600 °C. Figure 5.12 shows the SALDI MS spectrum of glucose. The spectrum is very clean with only alkali ions, Na^+ and K^+, of glucose. No protonated molecular peak was observed due to the low proton affinity of glucose. Arginine was also tested as shown in Fig. 5.13. Protonated arginine peak can be clearly identified with negligible background signal. Spectrum of crystal violet shows only the molecular ion peak without any other interference (Fig. 5.14). It has been reported that the carbon base SALDI materials, such as activated carbon, require the addition of a liquid such as glycerol; glycerol produces considerable background signal [10]. Signals can be obtained from small molecules using nanofibrous carbon substrate without adding any liquid. In addition, compared to the

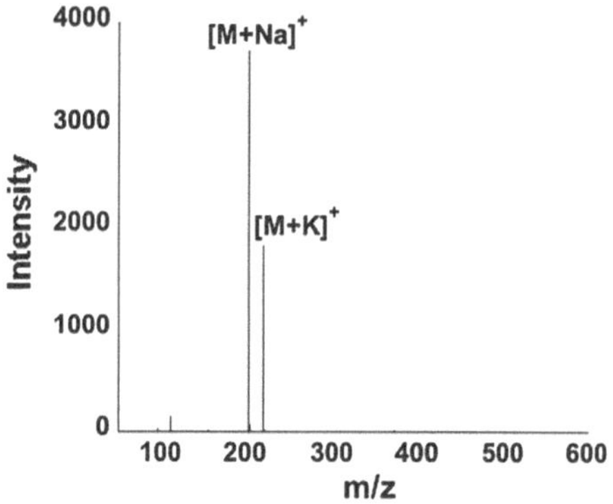

Fig. 5.12 SALDI-TOF mass spectrum of glucose (1 mg/mL) using the carbon substrate processed to 800 °C

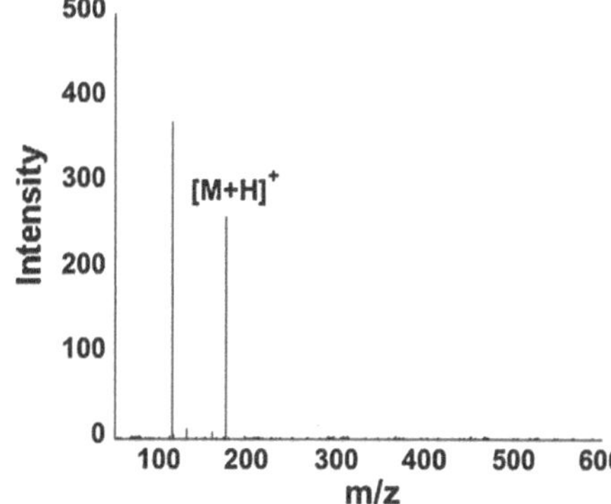

Fig. 5.13 SALDI-TOF mass spectrum of arginine (1 mg/mL) using the carbon substrate processed to 800 °C. The peak at m/z of 114 is from impurity. Similar spectrum were also obtained when ESI-TOF MS was used

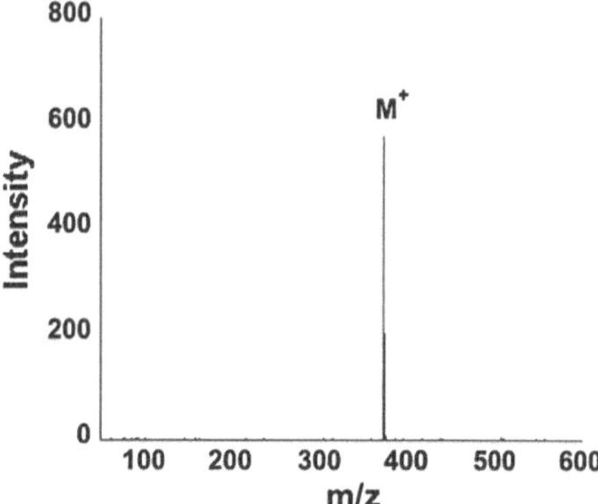

Fig. 5.14 SALDI-TOF mass spectrum of crystal violet (5 mg/mL) using the carbon substrate processed to 800 °C. Isotopic pattern was observed under high-resolution condition

SALDI MS using graphite which showed low mass interfering ion peaks from carbon clusters [45] the spectra of small molecules using carbon nanofibrous substrate showed no peaks from carbon clusters either. Therefore, the spectra of small molecules using carbon substrate processed to 800 °C have negligible background.

Glycerol was required to effectively dissipate excess heating energy for other SALDI materials, such as carbon, and avoid overheating of the analytes [9]. Heat transfer becomes much faster when the solid substrates are nanoscopic and under vacuum as occurs in the ion source [49, 50]. Therefore the heat dissipation rate is fast enough to prevent overheating and no addition of glycerol is needed. On the other hand, glycerol is required for nano-scale SALDI material, such as carbon nanotubes, to keep the carbon nanotubes to remain on the target without flying into the instrument [11]. Although the electrospun carbon nanofibers are small in diameter (~ 150 nm) they adhere on the target plate tightly due to their length and higher mass.

5.3.3 Polymeric Substrates

Many materials have been used as substrates for SALDI analyses because their ability to absorb the UV radiation at 337 nm, the wavelength of nitrogen laser commonly used in MALDI instrument. PAN, SU-8 and PVA have never been utilized for SALDI application perhaps due to their low UV absorptivity. Herein we have demonstrated for the first time that nanofibers of synthetic polymers with low UV absorption can be used for SALDI substrates as well. Synthetic polymer and small molecules were studied as analytes. Unlike carbon substrates no signal was observed from small molecules using polymeric substrates. However, signals from synthetic polymers were successfully obtained using polymeric substrates. Three concentrations of PS 4000 were tested, 25, 50 and 100 mg/mL, and were optimized for each substrate. Different laser input energies were used to obtain the optimal S/N ratio for PS 4000. The range of the laser input energy used in this work was similar to that used in previously reported result [44]. A typical plot of S/N versus laser input was shown in Fig. 5.15. As the laser energy increased an optimal S/N was acquired. A maximum laser power was found above which significant analyte fragmentation occurred.

Figures 5.16 and 5.17 shows the spectra using the optimal concentration and incident laser power. While no PS signal was observed from stainless steel substrates, high S/N was observed using nanofibrous PAN substrates (Fig. 5.16). The spectrum obtained has negligible background compared to those obtained on graphite (2 μm) and ZnO (20–200 nm), and the PS used has a higher M_w than those reported [9, 29]. PS with an $M_w = 10{,}000$ was also tested, but no signal was obtained. Signals were obtained from PS $M_w = 4{,}000$ with the PVA substrate, and both PS 4000 and PEG 3400 signals were obtained using nanofibrous SU-8 substrates, although the signals were poor. Figure 5.17 shows the spectrum obtained from PS using the SU-8 substrate. The laser power used for PS and PEG with the polymer-based substrates was slightly higher (~ 5 %) than that required for MALDI.

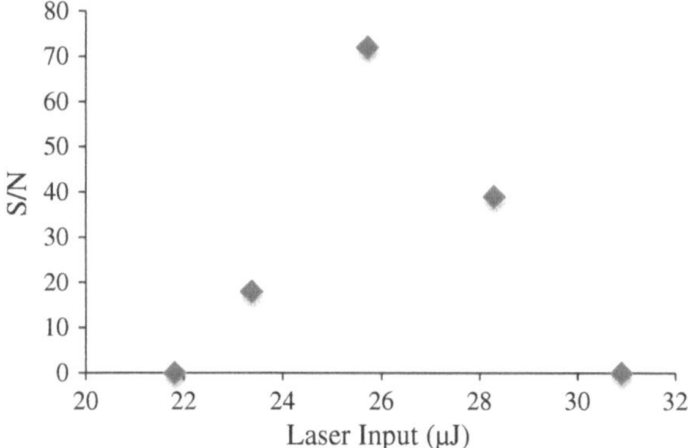

Fig. 5.15 Optimization of laser energy of PS 4000 (50 mg/mL) with PAN (10 %) substrate

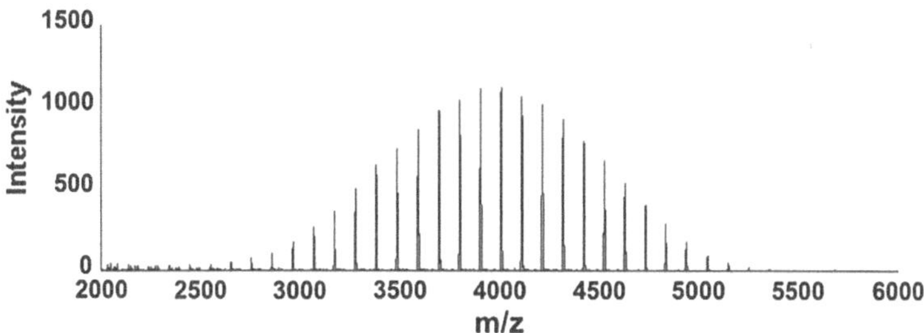

Fig. 5.16 SALDI-TOF mass spectrum of PS 4000 using the PAN substrate. PS concentration was 25 mg/mL

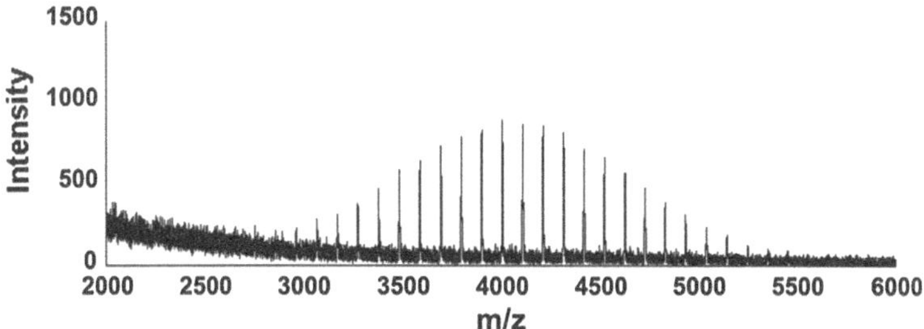

Fig. 5.17 SALDI-TOF mass spectrum of PS 4000 using the SU-8 substrate. PS concentration was 50 mg/mL

Similar to the SALDI results from the carbon substrates, sample molecules were more homogenously distributed than MALDI. Signals can be observed from every randomly chosen spot all over the sample similar to the carbon substrates. Table 5.4 compares the S/N ratio and its standard deviation of PS 4000 using the

Table 5.4 S/N of SALDI-TOF spectra with electrospun nanofibrous polymer substrates

PS 4000 (mg/mL)	25	50	100
PAN (RSD)	50 ± 10 (20 %)	40 ± 8 (20 %)	10 ± 10 (100 %)
SU-8 (RSD)	3.3 ± 0.4 (12 %)	3.6 ± 0.6 (17 %)	3.4 ± 0.7 (21 %)
MALDI (RSD)	40 ± 23 (60 %)		

S/N ratios were calculated based on ten spectra with 20 laser shots each

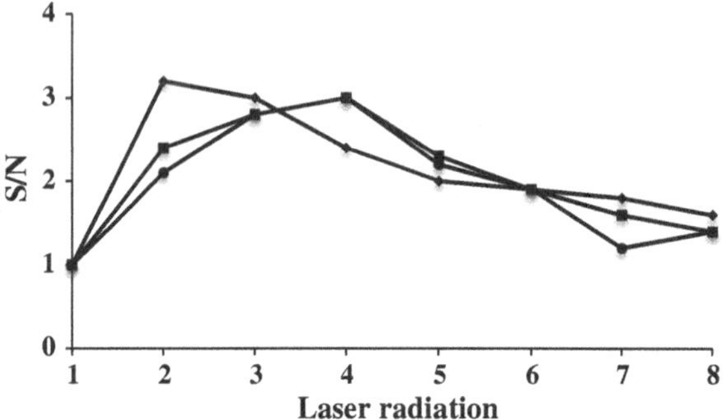

Fig. 5.18 S/N dependence with number of laser radiations on the same spot of the PS 4000 using PVA substrate. Each laser radiation contains 20 shots. (*filled circle*) 100 mg/mL, (*filled square*) 50 mg/mL, (*filled diamond*) 25 mg/mL

PAN and SU-8 substrates obtained from 10 random places. The relative standard deviation of PS 4000 using PAN substrate are about 20 % for 25 and 50 mg/mL compared to about 60 % using MALDI stainless steel target.

For the PS 4000 analysis using the PVA substrate, only background signal was obtained for the first 20 shots of laser radiation. Interestingly, when an additional 20 shots of laser radiation were applied to the same spot, the PS signal became apparent. The strongest signals were obtained between the 2nd to 4th set of laser irradiation (20 shots each) to the same region, depending on the concentration of the PS (Fig. 5.18). For the lowest concentration, the strongest signal was observed from the 2nd set of 20-shots of laser radiation, and for the two higher concentrations, the strongest signals were obtained from the 3rd or 4th set of 20-shots of laser radiation. This was only observed when using PVA substrate. The reason was not clear. It may be related to polymer-polymer interactions or partitioning of the low molecular weight polymer into the fibers.

Background signal from PAN, SU-8, and PVA was not observed in any of these measurements.

Dynamic range of the ionization method is of considerable importance when determining the correct molecular weight of a polymer by mass spectrometry. Using a method described by Zhu et al. [51] initial studies of the dynamic range for polymer molecular weight determinations were undertaken. A polymer blend of 5 % of a low molecular weight polymer is mixed with 95 % of a high molecular weight polymer; the low molecular weight polymer should be detectable in the blend if the method is sensitive and provides good dynamic range for polymer analysis. Figure 5.19 shows the spectrum of the mixture of PS 4000 and PS 2000.

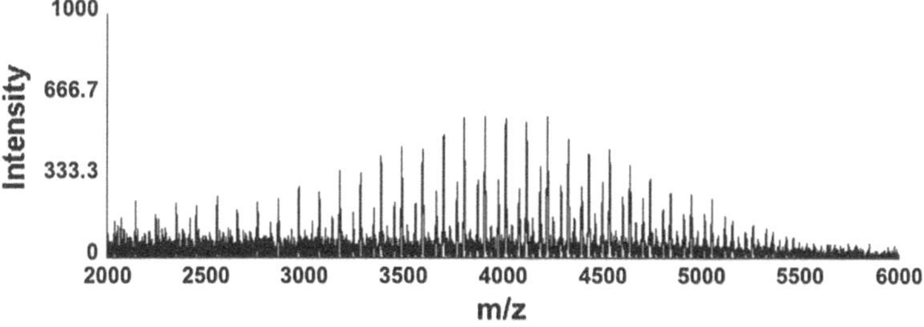

Fig. 5.19 SALDI-TOF mass spectrum of 5 % PS 2000 mixed with PS 4000 (mol %)

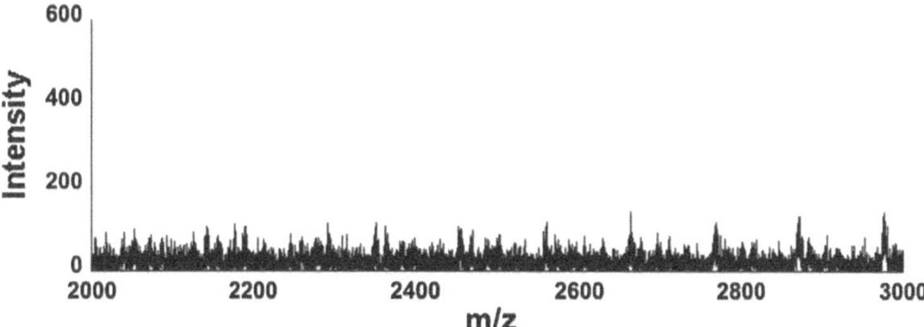

Fig. 5.20 The expanded low mass region SALDI-TOF mass spectrum of PS 4000

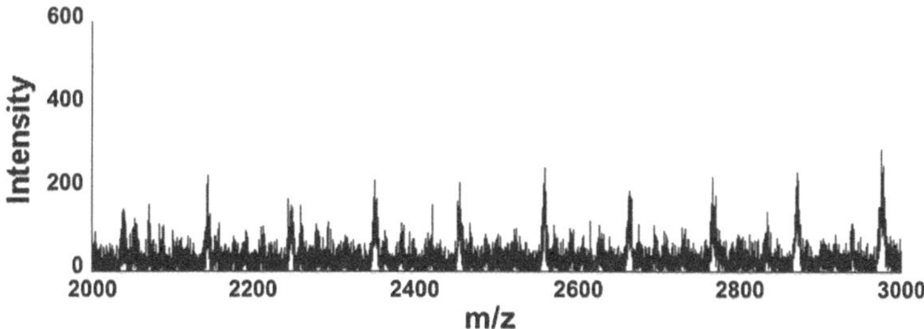

Fig. 5.21 The expanded low mass region of SALDI-TOF mass spectrum of 5 % PS 2000 mixed with PS 4000 (mol %)

The low mass tails of the spectra of PS 4000 and PS 4000/PS2000 blend using the nanofibrous PAN substrate showed that PS 2000 at the low mass tail was clearly observed (Figs. 5.20 and 5.21), which demonstrated that good dynamic range was achieved using the nanofibrous PAN substrate.

To investigate the possible importance of fiber morphology and pore structure to the observed signal intensity (S/N), PAN nanofibers with different diameters and pore sizes were prepared. By increasing the PAN concentration to 12.5 % for electrospinning, larger fibers were prepared [52]. The fiber diameter (880 nm) was much larger than the fibers obtained from 10 % PAN solution, and thus the surface

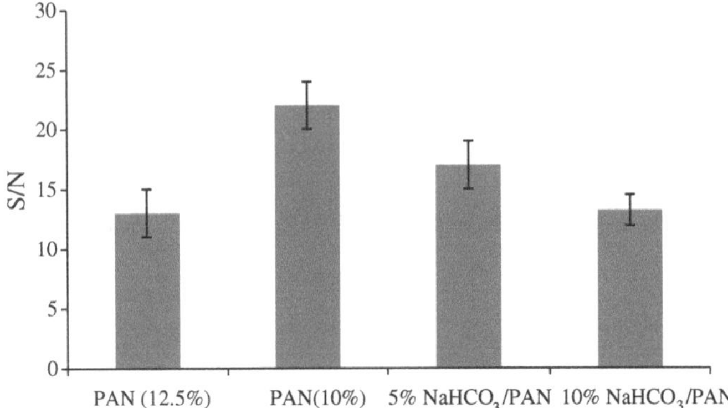

Fig. 5.22 S/N of PS 4000 (50 mg/mL) on PAN substrate with different diameters, pore sizes and pore density

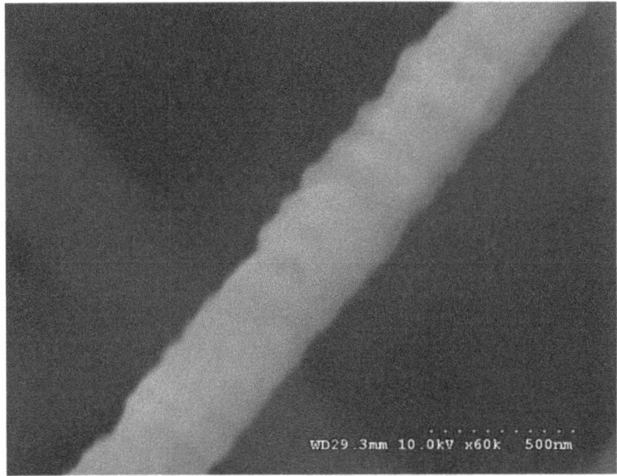

Fig. 5.23 SEM image of porous PAN prepared by adding 10 % $NaHCO_3$

area was lower than the substrate prepared from 10 % PAN solution. The S/N using the substrate made from 12.5 % PAN was lower than that observed using smaller diameter fibers (Fig. 5.22). Similar results were obtained by Kuznetsova et al. using carbon materials with different surface areas [53].

Larger pores were incorporated into the PAN fibers by adding $NaHCO_3$ to the PAN solution and pores were formed when $NaHCO_3$ was removed after electrospinning using hydrochloric acid [54]. The average pore diameter was about 10–50 nm; with increased concentration of $NaHCO_3$ in the electrospun fiber the pore density also increases [54]. Two concentration of $NaHCO_3$ were used here, 5 and 10 %. The obtained PAN fibers showed porous morphology under SEM (Fig. 5.23) and remained similar fiber diameter to the pure PAN (Table 5.2). The S/N of PS 4000 of PAN substrates with different pore morphologies is shown in Fig. 5.22. It is clear that the pore morphology has significant impact of S/N. PAN fibers with larger pores showed lower S/N than the pure PAN substrate with similar fiber diameter (460 nm). The possible reason is due to the pore size (10–50 nm) is too large and they do not retain the polymer analyte molecules as well as the micropores in pure PAN (<2 nm) [55].

Unlike other SALDI polymeric substrates which have intensive UV absorption [22, 29], the polymers used in this study included PAN, SU-8 and PVA all have low absorption at the laser wavelength, 337 nm. However, PS signals were observed on all of these polymer substrates. Interestingly, the S/N ratio of PS with the PAN substrate was about ten times higher than the S/N ratio with the SU-8 substrate although the fibers have similar diameters (Figs. 5.16 and 5.17), which indicates that the ionization efficiency does not completely rely on the UV absorption extinction coefficient or the size of the materials. Unlike the other SALDI materials, the nanofibrous polymers have a unique microporous structure which is a result of the interstices between the polymer chains. Consequently, the polymeric analyte molecules can partition into the polymer nanofibers instead of exclusively adsorbing on the surface. We believe that the partitioning process is crucial for the success of SALDI analyses using the nanofibrous substrates. Although the mechanism of SALDI has not been fully described yet, it is believed that the thermal desorption/ionization plays a prominent role [17, 40, 45]. Polymers are known to be poor heat conductors. Upon laser radiation, the temperature of the analyte molecule, which is localized in the polymer fiber, increases substantially without losing heat to the thermally insulating surroundings. Compared to PAN, PVA and SU-8 substrates are cross-linked, which may lead to a lower microporosity. Therefore, the partitioning could be more difficult for the SU-8 and PVA substrates, which explains their poorer SALDI behavior, compared to the PAN substrate.

5.3.4 ME-SALDI Analysis with Electrospun Nanofibrous Substrates

ME-SALDI analysis with electrospun nanofibrous substrates was performed and high sensitivity was achieved. The diameter of the sample spot prepared by the dried droplet method was 1 to 5 mm. The sensitivity of MALDI is limited by the small sampling size which is the spot size of the laser radiation, 50 to 200 μm. Hence, only a small fraction of the sample is exposed to the laser and is analyzed [56]. With high surface area, the electrospun nanofibrous substrates may be able to trap more analyte molecules and thus enhance the sensitivity.

Laser input energy has significant impact on the intensity and mass resolution. Excessive laser energy can cause secondary processes in the matrix plume above the target plate, which may lead to a wider energy distribution of the ions with the same mass and as a result the resolution decreases. Herein we systematic studied the S/N ratios and the resolution under different laser input energies for

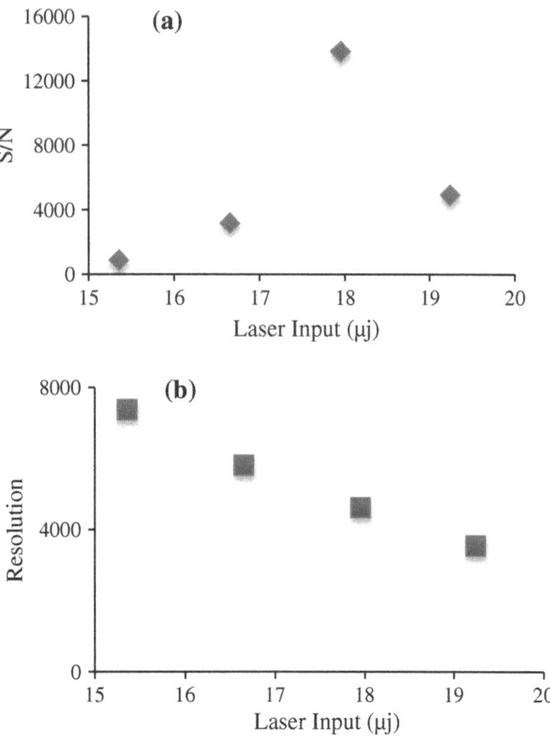

Fig. 5.24 **a** Optimization of S/N of angiotensin I (2 mg/mL) using CHCA (2 × 10 mg/mL) as matrix with carbon substrate processed to 800 °C and **b** the resolution change versus the laser input power

ME-SALDI. Figures 5.24 and 5.25 shows the optimization of S/N and resolution with different laser input energies for angiotensin I as the analyte using carbon substrate processed to 800 °C and PVA substrate. The S/N increased when the laser input power increased until reaching the optimal laser energy. Then the S/N decreased due to increased noise with further increases of laser energy. The resolution decreased slightly when higher laser energy was used for carbon substrate while for PVA substrate the resolution remained fairly stable in the range of the laser energy that applied. All the other ME-SALDI analysis conditions were optimized using the similar method.

When different substrates were used, the optimal matrix/analyte ratios changed. The matrix/analyte ratios for analytes at three different concentrations were optimized on all substrates including stainless steel (Table 5.1). The applied volume of sample and matrix solution was 0.1 μL. 20 shots were used to acquire each spectrum. “Sweet spot” were searched all over the sample spot to obtain the optimum signal S/N rations. Five replicates were performed for each sample. Since the standard deviation listed in Table 5.1 was calculated from five spectra obtained only from “sweet spots” by searching all over the sample, the standard deviation cannot be used to represent the shot-to-shot reproducibility like that discussed in the SALDI section.

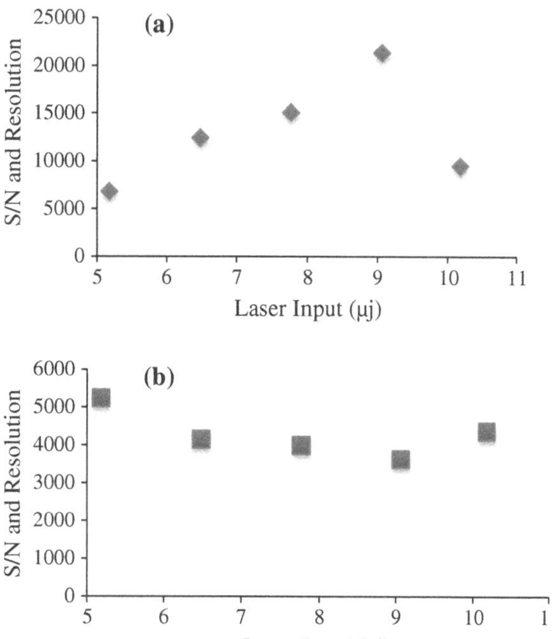

Fig. 5.25 **a** Optimization of S/N of angiotensin I (2 mg/mL) using CHCA (2 × 10 mg/mL) as matrix with PVA substrate. **b** The resolution change versus the laser input power

For PEG 3400 as the analyte with DHB as the matrix using SU-8 and PAN substrates, considerably enhanced S/N ratios were observed compared to those observed using the stainless steel substrate (Fig. 5.26). For example, the S/N of PEG (20 mg/mL) with the SU-8 substrate is about 2.6 times larger than the S/N obtained with the stainless steel substrate. For the PS 4000 with dithranol as the matrix, the PAN and PVA substrates showed enhanced S/N ratios compared to the stainless steel substrate (2.0 times larger for 100 and 50 mg/mL, respectively) (Fig. 5.27). Although no signal from angiotensin I was observed from electrospun SALDI, S/N was improved on electrospun substrates under ME-SALDI condition. For angiotensin I (0.5 mg/mL), the carbon substrate processed to 800 °C showed significantly enhanced S/N (~4,300) compared to the S/N of the stainless steel substrate (~1,500). The PVA substrate and carbon substrate processed to 600 °C showed improved S/N ratios as well for angiotensin I detection (Fig. 5.28). Compared to the stainless steel substrate, all of the nanofibrous substrates yielded ME-SALDI enhanced S/N ratios. ME-SALDI analyses using the electrospun nanofibrous substrates showed similar resolution to that of stainless steel substrate.

Clean background has been observed in previous ME-SALDI study [55]. Electrospun nanofibrous carbon and PVA substrates showed the similar matrix suppression effect. As the spectra shown in Figs. 5.29, 5.30 and 5.31 different concentrations, 2, 1 and 0.5 mg/mL, of angiotensin I with the same matrix/analyte ratio were spotted on the carbon substrate processed to 800 °C. As the concentration of analyte decreased the background signals, especially in the low mass

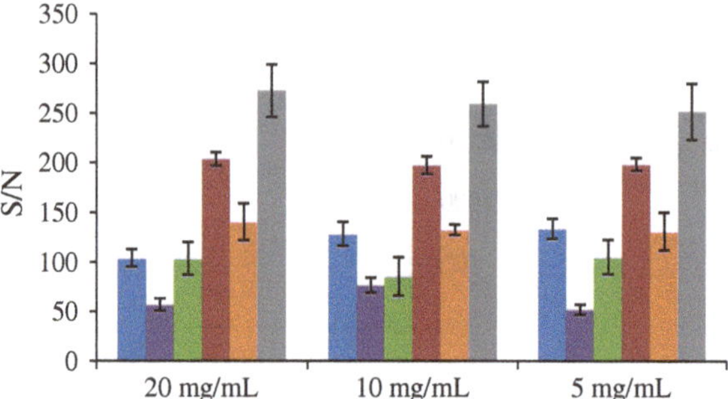

Fig. 5.26 S/N ratios of PEG 3400 on different substrates. (*blue bar*) stainless steel, (*green bar*) carbon processed to 800 °C, (*violet bar*) carbon processed to 600 °C, (*brown bar*) PAN, (*orange bar*) PVA, (*grey bar*) SU-8. S/N ratios calculated from five spectra with 20 shots each. The error bars show the standard deviations. DHB was used as matrix. The matrix/analyte ratios and laser power were optimized

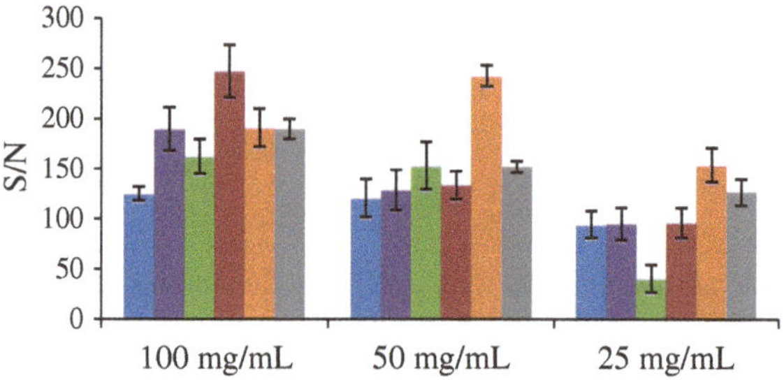

Fig. 5.27 S/N ratios of PS 4000 on different substrates. (*blue bar*) stainless steel, (*green bar*) carbon processed to 800 °C, (*violet bar*) carbon processed to 600 °C, (*brown bar*) PAN, (*orange bar*) PVA, (*grey bar*) SU-8. S/N ratios calculated from five spectra with 20 shots each. The error bars show the standard deviations. Dithranol was used as matrix. The matrix/analyte ratios and laser power were optimized

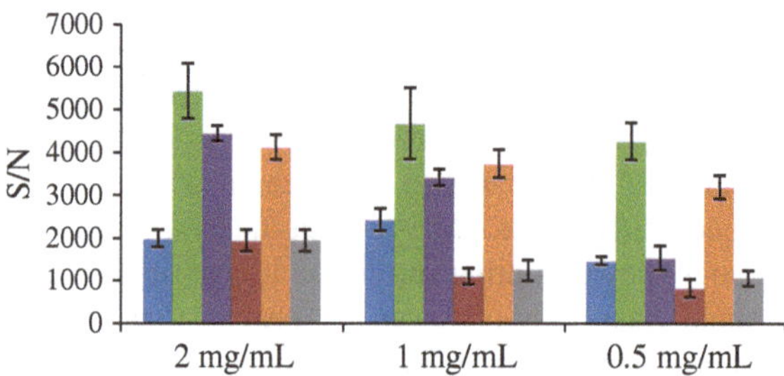

Fig. 5.28 S/N ratios of angiotensin I on different substrates. (*blue bar*) stainless steel, (*green bar*) carbon processed to 800 °C, (*violet bar*) carbon processed to 600 °C, (*brown bar*) PAN, (*orange bar*) PVA, (*grey bar*) SU-8. S/N ratios calculated from five spectra with 20 shots each. The error bars show the standard deviations. CHCA was used as matrix. The matrix/analyte ratios and laser power were optimized

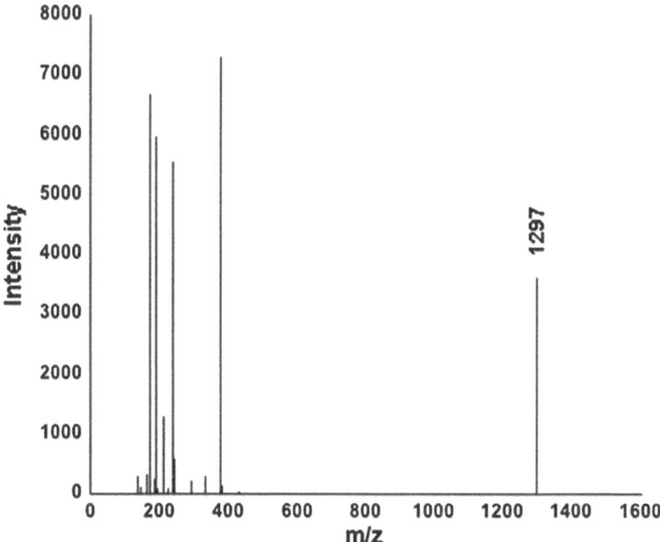

Fig. 5.29 ME-SALDI spectrum of angiotensin I using carbon substrate processed to 800 °C with concentrations of 2 mg/mL. The matrix/analyte ratio is 1:10 (w/w)

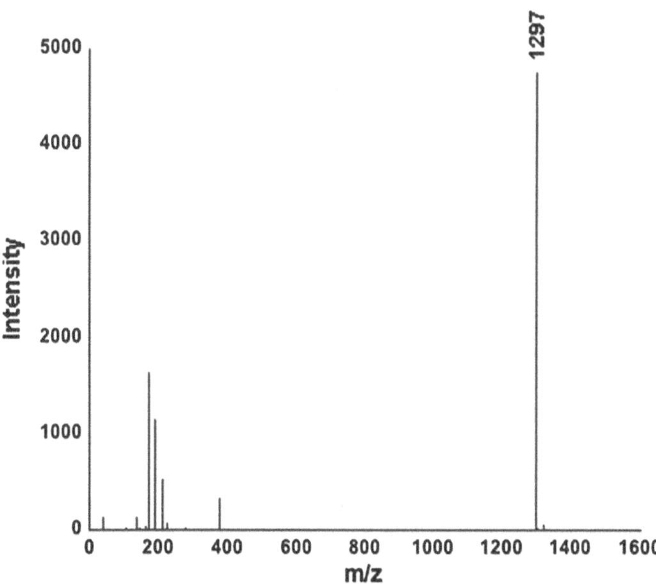

Fig. 5.30 ME-SALDI spectrum of angiotensin I using carbon substrate processed to 800 °C with concentrations of 1 mg/mL. The matrix/analyte ratio is 1:10 (w/w)

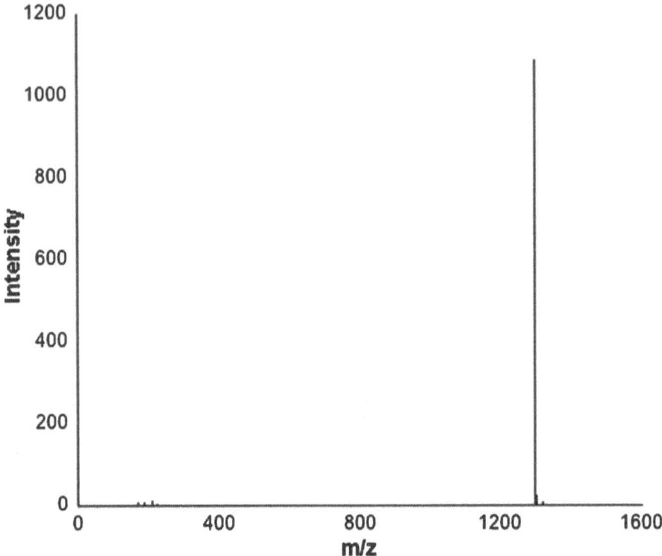

Fig. 5.31 ME-SALDI spectrum of angiotensin I using carbon substrate processed to 800 °C with concentrations of 0.5 mg/mL. The matrix/analyte ratio is 1:10 (w/w)

region, became lower. Although the S/N was smaller the cleanest background was obtained with 0.5 mg/mL angiotensin I (Fig. 5.31) which is better than using stainless steel substrate (Fig. 5.32). The PVA substrate also showed low background signal for angiotensin I analysis in the low mass region (Fig. 5.33). It is interesting that the laser energy needed for angiotensin I was significantly lower when using PVA substrate than that using carbon or stainless steel substrate as shown in Fig. 5.25. This might be the reason that PVA yields cleaner background in the low mass region than carbon and stainless steel substrates.

The morphology of all the nanofibrous substrates was investigated after they were prepared with the analyte and matrix solutions using SEM. All of the substrates maintained their nanofibrous morphology after applying the sample solutions with the exception of PVA which lost part of its nanofibrous morphology after applying angiotensin I and CHCA solutions. The cross-linked PVA might be unstable in the acidic solution which is required to stabilize CHCA matrix. Nevertheless, the PVA substrate still showed enhanced sensitivity for angiotensin I compared to other reported SALDI methods [46]. Figure 5.34 shows the spectrum of 800 attmol angiotensin I, which is smaller than the detection limit of MALDI-MS; the S/N of the angiotensin I peak is 18. The detection limit was studied using angiotensin I with the carbon substrate processed to 800 °C. When 400 attmol angiotensin I was applied, a S/N of 3.2 was obtained (Fig. 5.35). Hence we can estimate the detection limit to be 400 attmol, which is smaller compared to the low fmol detection limit of a peptide with MALDI/SALDI [57] using the dried droplet sample preparation method.

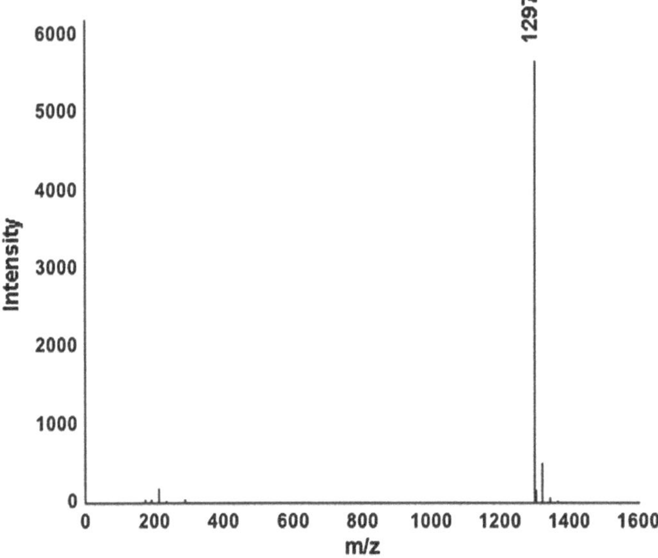

Fig. 5.32 ME-SALDI spectrum of angiotensin I using stainless steel substrate with concentration of 0.5 mg/mL. The matrix/analyte ratio is 1:10 (w/w)

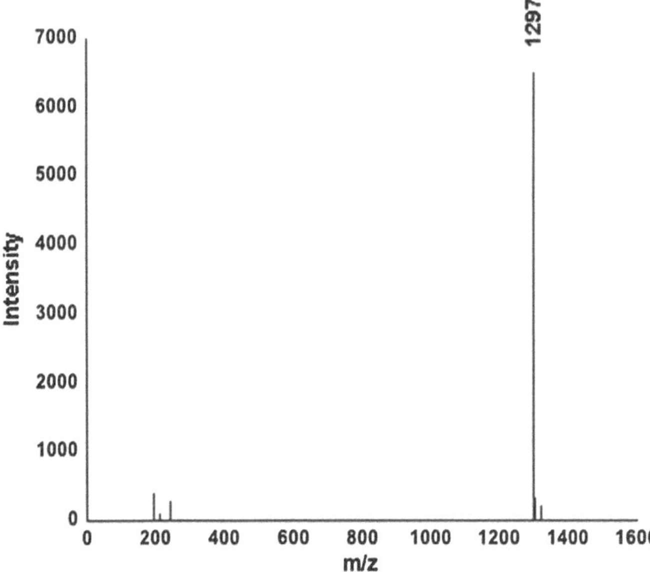

Fig. 5.33 ME-SALDI spectrum of angiotensin I using PVA substrate with concentration of 1 mg/mL. The matrix/analyte ratio is 1:10 (w/w)

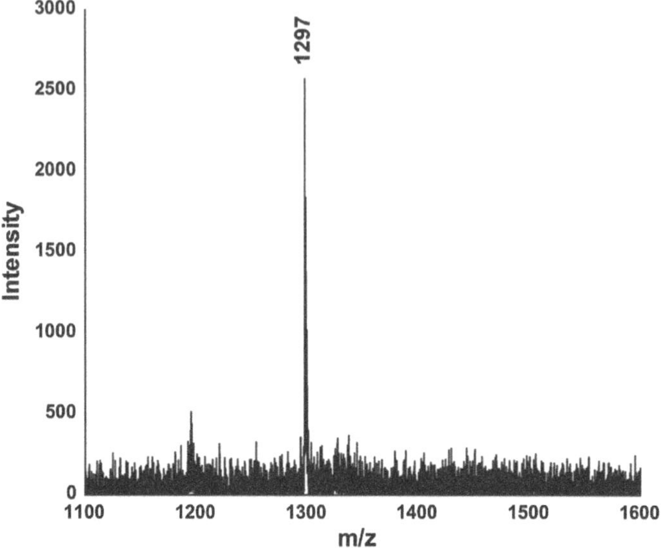

Fig. 5.34 ME-SALDI-TOF mass spectrum of angiotensin I with the PVA substrate. 800 attmol of angiotensin I was applied and the CHCA concentration was 0.1 mg/mL

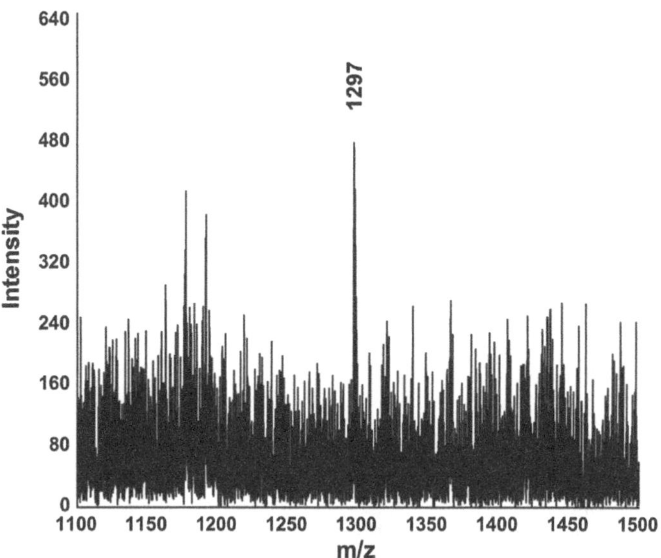

Fig. 5.35 ME-SALDI-TOF spectrum of angiotensin I with the carbon substrate processed to 800 °C

The amount of angiotensin I applied was 400 attmol and the CHCA concentration was 0.05 mg/mL.

5.4 Conclusion

Electrospun PAN, PVA, and SU-8 nanofibers as well as carbon nanofibers pyrolyzed from SU-8 nanofibers as substrates for SALDI and ME-SALDI shows considerable potential. First, the sample preparation for SALDI is simple and no organic matrix is needed. Secondly, Electrospun carbon nanofibrous substrates for SALDI showed excellent ionization efficiency for large molecules. Therefore they provide a convenient SALDI method for high molecular weight synthetic polymers. Thirdly, electrospun nanofibrous substrates are advantageous for small molecule analysis generating negligible interfering background signals. Finally, as ME-SALDI substrates electrospun nanofibers, especially carbon and PVA substrates, improved sensitivity compared to both conventional MALDI and other SALDI. They serve as a sensitive method for analysis requiring extreme low detection limit. Therefore, electrospun nanofibrous SALDI analysis provides a simple method for analytes with large molecular weight. ME-SALDI analysis using electrospun nanofibers can be used for analysis of samples with low concentration because this method provides improved detection limit. Many other polymers can be electrospun with a variety of fiber sizes and morphologies, which may increase the breath of molecules that can be detected using laser desorption ionization techniques.

References

1. Tanaka K, Waki H, Ido Y, Akita S, Yoshida Y, Yohida T (1988) Rapid Commun Mass Spectrom 2:151
2. Karas M, Bachmann D, Hillenkamp F (1985) Anal Chem 57:2935
3. Vorm O, Roepstorff P, Mann M (1994) Anal Chem 66:3281
4. Axelsson J, Hoberg A-M, Waterson C, Myatt P, Shield GL, Varney J, Haddleton DM, Derrick PJ (1997) Rapid Commun Mass Spectrom 11:209
5. Jaskolla TW, Karas M, Roth U, Steinert K, Menzel C, Reihs K (2009) J Am Soc Mass Spectrom 20:1104
6. McCombie G, Knochenmuss R (2006) J Am Soc Mass Spectrom 17:737
7. Umemura H, Kodera Y, Nomura F (2010) Clin Chim Acta 411:2109
8. Sunner J, Dratz E, Chen Y-C (1995) Anal Chem 67:4335
9. Dale MJ, Knochenmuss R, Zenobi R (1996) Anal Chem 68:3321
10. Han M, Sunner J (2000) J Am Soc Mass Spectrom 11:644
11. Pan C, Xu S, Zou H, Guo Z, Zhang Y, Guo B (2005) J Am Soc Mass Spectrom 16:263
12. Wei J, Buriak JM, Siuzdak G (1999) Nature 399:243
13. Greving MP, Patti GJ, Siuzdak G (2011) Anal Chem 83:2
14. Nayak R, Knapp DR (2010) Anal Chem 82:7772

15. Kawasaki H, Yao T, Suganuma T, Okumura K, Iwaki Y, Yonezawa T, Kikuchi T, Arakawa R (2010) Chem Eur J 16:10832
16. Chen WT, Chiang CK, Lee CH, Chang HT (2012) Anal Chem 84(4):1924
17. Tang HW, Ng KM, Lu W, Che CM (2009) Anal Chem 81:4720
18. Alimpiev S, Nikiforov S, Karavanskii V, Minton T, Sunner J (1891) J Chem Phys 2001:115
19. Kruse RA, Li X, Bohn PW, Sweedler JV (2001) Anal Chem 73:3639
20. Schuerenberg M, Dreisewerd K, Hillenkamp F (1999) Anal Chem 71:221
21. Zhang J, Li Z, Zhang C, Feng B, Zhou Z, Bai Y, Liu H (2012) Anal Chem 84:3296
22. Woldegiorgis A, Lowenhielm P, Bjork A, Roeraade J (2004) Rapid Commun Mass Spectrom 18:2904
23. Woldegiorgis A, Von KF, Dahlstedt E, Hellberg J, Brinck T, Roeraade J (2004) Rapid Commun Mass Spectrom 18:841
24. Hua L, Low TY, Meng W, Chan-Park MB, Sze SK (2007) Analyst 132:1223
25. Peterson DS, Luo Q, Hilder EF, Svec F, Frechet JMJ (2004) Rapid Commun Mass Spectrom 18:1504
26. Shen Z, Thomas JJ, Averbuj C, Bro KM, Engelhard M, Crowell JE, Finn MG, Siuzdak G (2001) Anal Chem 73:612
27. Kim HJ, Lee J-K, Park S-J, Ro HW, Yoo DY, Yoon DY (2000) Anal Chem 72:5673
28. Kim J, Kang W (2000) Bull Korean Chem Soc 21:401
29. Watanabe T, Kawasaki H, Yonezawa T, Arakawa R (2008) J Mass Spectrom 43:1063
30. Chen Z, Geng Z, Shao D, Mei Y, Wang Z (2009) Anal Chem 81:7625
31. Kawasaki H, Yonezawa T, Watanabe T, Arakawa R (2007) J Phys Chem C 111:16278
32. Chiang C, Yang Z, Lin Y, Chen W, Lin H, Chang H (2010) Anal Chem 82:4543
33. Liu Q, Xiao Y, Pagan-Miranda C, Chiu YM, He L (2009) J Am Soc Mass Spectrom 20:80
34. Liu Q, He L (2009) J Am Soc Mass Spectrom 20:2229
35. Lu T, Olesik SV (2013) J Chromatogr B Anal Technol Biomed Life Sci 912:98
36. Clark JE, Olesik SV (2009) Anal Chem 81:4121
37. Ma G, Yang D, Nie J (2009) Polym Adv Technol 20:147
38. Steach JK, Clark JE, Olesik SV (2010) J Appl Polym Sci 118:405
39. Clark JE, Olesik SV (2010) J Chromatogr A 1217:4655
40. Meier MAR, Schubert US (2003) Rapid Commun Mass Spectrom 17:713
41. Kempka M, Sjodahl J, Bjork A, Roeraade J (2004) Rapid Commun Mass Spectrom 18:1208
42. Steach JK, Clark JE, Olesik SV (2010) J Appl Polym Sci 118:405
43. Mizuno K, Ishii J, Kishida H, Hayamizu Y, Yasuda S, Futaba DN, Yumura M, Hata K (2009) PNAS 106:6044
44. Mattern DE, Hercules DM (1985) Anal Chem 57:2041
45. Pan C, Xu S, Hu L, Su X, Qu J, Zou H, Guo Z, Zhang Y, Guo B (2005) J Am Soc Mass Spectrom 16:883
46. Buryak AK, Serdyuk TM (2011) Prot Met Phys Chem Surf 47:911
47. Schriemer DC, Li L (1996) Anal Chem 68:2721
48. Li L (2010) MALDI mass spectrometry for synthetic polymer analysis. Wiley, Hoboken
49. Chen G, Shakouri A (2002) ASME J Heat Transf 124:242
50. Cheng C, Fan W, Cao J, Ryu S, Ji J, Grigoropoulos CP, Wu J (2011) ACS Nano 5:10102
51. Zhu H, Yalcin T, Li L (1998) J Am Soc Mass Spectrom 9:275
52. He J, Wan Y, Yu J (2008) Fibers Polym 9:140
53. Kuznetsova ES, Buryak AK (2011) Prot Met Phys Chem Surf 47:713
54. Ma G, Yang D, Nie J (2009) Polym Adv Technol 20:147
55. Jang J, Bae J (2005) J Chem Commun 9:1200
56. Wei H, Nolkrantz K, Powell DH, Woods JH, Ko M, Kennedy RT (2004) Rapid Commun Mass Spectrom 18:1193
57. Kuzema PA (2011) J Anal Chem 66:1227

Chapter 6
Synthesis of Fluorescent Carbon

Abstract Water insoluble fluorescent carbon particles were prepared by covalently attaching a BODIPY fluorophore to the surface. The fluorescence wavelength is not dependent on excitation wavelengths. Preliminary results on using the fluorescent carbon particles for the study of micro/nanofluidics are included.

6.1 Introduction

Fluorescent carbon particles may be of value for many fields of science. Carbon materials are chemically stable and nontoxic. Therefore, carbon materials have been widely-used as environment friendly materials. Due to this reason carbon materials with luminescence have great potential in bioimaging and biolabeling [1]. In addition, carbon particles with fluorescent emission are of great interest in visualization of the movement of carbon particles. Polymers and metals with photoluminescence have been previously used as tracers to study the fluidics in electrokinetically driven flows [2]. On the other hand, carbon with its unique density and dielectric constant is less studied. Therefore, it is important to study the movement of carbon particles as well.

Preparation of fluorescent carbon nanoparticles with small diameter, <5 nm, has been reported [1]. Due to the small size comparable to the quantum dots (QDs), they are also called carbon dots (C-dots). C-dots passivated with polymers or organic molecules usually exhibit high fluorescence quantum yield [3]. This unusual luminescence character of C-dots is not fully understood yet. The energy-trapping sites on the surface and their radiative recombination have been suggested to be the reason for the luminescence [3]. The preparation of C-dots is time-consuming and requires an expensive experimental apparatus, such as arc-discharge [4] and laser ablation, which limits their practical applications [3]. Although the C-dots show strong fluorescence they are usually soluble in water and cannot be used as tracers. Another way to prepare fluorescent carbon is using strong oxidizing reagents to oxidize the surface of carbon particles [5]. After this treatment, the carbon becomes fluorescent. The fluorescence is believed to be a result of the surface oxidation and

T. Lu, *Nanomaterials for Liquid Chromatography and Laser Desorption/Ionization Mass Spectrometry*, Springer Theses,
DOI: 10.1007/978-3-319-07749-9_6,

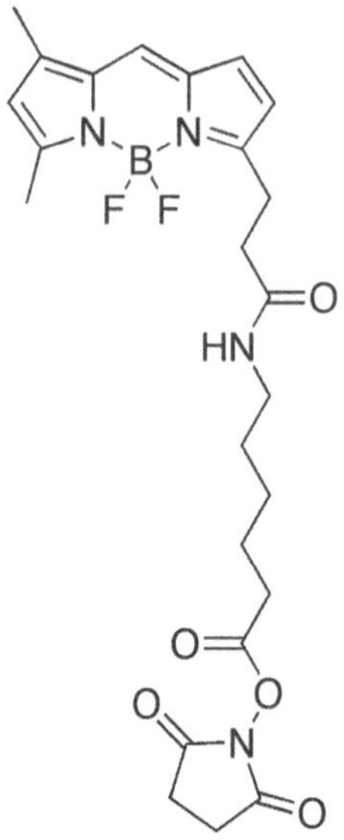

Fig. 6.1 The structure of BODIPY FL-X, SE, (6-((4, 4-difluoro-5, 7-dimethyl-4-bora-3a, 4a-diaza-*s*-indacene-3-propionyl)amino)hexanoic acid, succinimidyl ester)

nitrogen and oxygen doping by adding a conjugated π system that can generate photoluminescence [6]. The preparation is simple and larger carbon particles can be used. The drawback of this method is, however, after the oxidizing treatment, the carbon particles become water soluble due to the formation of hydrophilic functional groups, $-COOH$, $-NO_2$ and $-OH$, on the surface [5].

It is important to note that carbon acts as a quencher for the fluorescence of organic dyes [7]. To ensure a high fluorescence quantum yield of the prepared carbon particles, a spacer that separates the dye molecule from the carbon surface is required. The quenching is through Förster resonance energy transfer process. The relationship between the distance and the quantum yield of the quenching is: [8]

$$E = 1\Big/\left[1 + (r/R_0)^6\right], \tag{6.1}$$

where r is the distance between the fluorophore and quenching molecule and R_0 is the Förster distance. The Förster distance is the distance between the donor and the acceptor at which the energy transfer efficiency is 50 % [8]. From this equation we know that by increasing the distance between the fluorophore and carbon surface, the quenching should be significantly decreased.

In this chapter, we choose 4, 4-difluoro-5, 7-dimethyl-4-bora-3a, 4a-diaza-*s*-indacene (BODIPY) as a fluorophore. The structure of the dye used in this work is shown in Fig. 6.1. The fluorescence quenching process has been studied using BODIPY and peptides as quenchers. The BODIPY and the peptides were connected using alkyl chains with different lengths. The fluorescence can be quenched by the peptides through a photoinduced electron transfer process [9]. The results show that the BODIPY fluorescence quenching yield can be greatly decreased by increasing the distance from the quencher [10].

We prepared fluorescent carbon by chemically attaching fluorophores on the carbon particle surface. The prepared fluorescent carbon particles show great advantages for the visualization of the movement of the carbon particles in the fluidic devices. First, the fluorescent carbon particles exhibit bright fluorescence and low solubility in water. Secondly, the fluorescence quantum yield can be easily tuned by the linker distance and the choice of the molecular fluorophore. Thirdly, the carbon particle size can be easily changed. In this chapter, we focus on the synthesis of fluorescent carbon particles with a size of 5 μm and 100 nm by covalently attaching a BODIPY fluorophore.

Alkyl-substituted BODIPY exhibits a fluorescence quantum yield close to 1 [11]. The length of the alkyl chain that is not conjugated does not change the fluorescence emission significantly [12]. We apply a silane coupling reaction to immobilize a silane reagent, with a linker functionalized with the primary amine, on the pre-hydroxylated carbon surface. At the last step, BODIPY was attached to the amine functionalized surface through the formation of amide bond. BODIPY is separated from the carbon surface by a 22-atom spacer. Water insoluble carbon particles with strong fluorescence were successfully prepared.

6.2 Experiment

6.2.1 Reagents and Instruments

Potassium permanganate, tetrapropyl ammonium bromide (TPABr), hydroquinone, sodium carbonate, sodium bicarbonate, rhodamine B and activated carbon were purchased from Sigma Aldrich (St. Louis, MO). *N*-(6-aminohexyl)aminomethyl triethoxysilane (AHAMTES) was purchased from Gelest (Morrisville, PA). BODIPY FL-X succinimidyl ester was purchased from Invitrogen (Carlsbad, CA). Toluene, hydrochloric acid, methylene chloride, dimethylformamide (DMF), acetic acid and ethanol were purchased from Fisher Scientific.

6.2.2 Synthesis of Fluorescent Carbon

Fluorescent carbon nanoparticles were prepared by attaching a fluorescent dye on to the carbon surface. The synthetic scheme is shown in Fig. 6.2.

Carbon surface hydroxylation The carbon surface hydroxylation was performed following a literature procedure [13]. Both activated carbon (60–100 nm) and Hypercarb (5 μm) were used for the preparation of the fluorescent carbon particles. Typically 0.1 g of carbon and 100 mL of methylene chloride were added into a flask and sonicated for 10 min to disperse the carbon particles. Then a $KMnO_4$ solution (0.25 g in 12.5 mL H_2O/acetic acid, 2:3) and 0.12 g TPABr were added. The reaction mixture was stirred for 24 h at room temperature. Then the mixture was

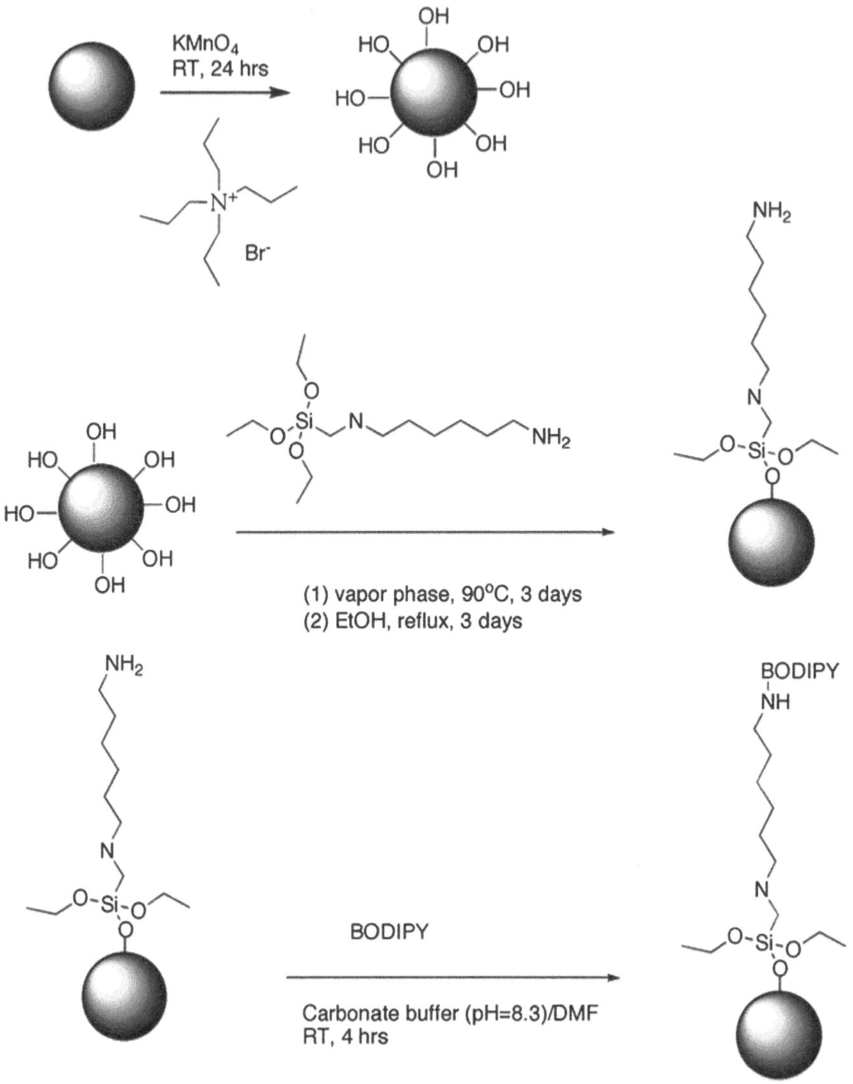

Fig. 6.2 Synthetic scheme of fluorescent carbon

washed with HCl and methanol. The liquid portion was removed by centrifugation and the carbon particles were dried under vacuum. The yield of the reaction is 80 %.

Silane coupling-vapor phase AHAMETES is a silane-coupling reagent that can react with the hydroxyl groups on the carbon surface. It has an amine group in the other end of the alkyl chain that can react with BODIPY. The vapor phase reaction was conducted following a literature procedure [14]. About 1 mL of AHAMTES was added into a 10 mL flask. The hydroxylated carbon particles were added into a small vial (1.5 mL) and then the small vial containing the carbon particles was put

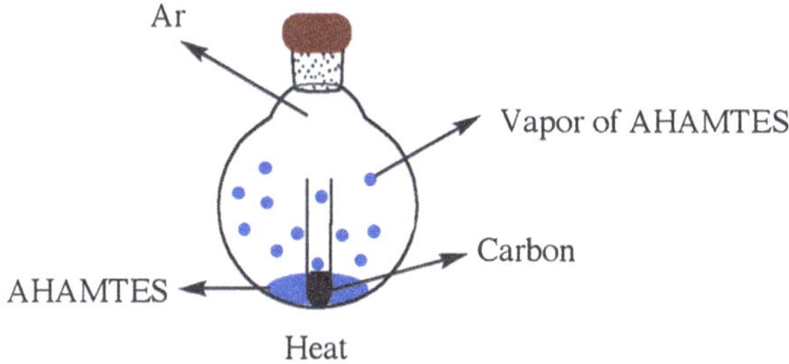

Fig. 6.3 Reaction setup of vapor phase silane coupling

into the flask containing the silane reagent (Fig. 6.3). The liquid and carbon need to be separated completely to avoid solution phase reaction. Then the system was sealed and purged with Ar to remove water, and heated to 90 °C using an oil bath for 3 days. The carbon particles were used for further modification without purification.

Silane coupling-solution phase [15] The hydroxylated carbon particles (50 mg) were dispersed in ethanol (50 mL) through sonication for 20 min. 0.1 mL of AHAMTES was added to the reaction mixture. The reaction mixture was heated to reflux for 3 days. Then the particles were washed with ethanol thoroughly to remove the unreacted AHAMTES. The particles were then dried in vacuum.

BODIPY attachment 50 mg silanized carbon particles were dispersed in 0.5 mL $NaHCO_3$ buffer solution (0.1 M, pH = 8.3). BODIPY (5 mg) was dissolved in DMF (0.5 mL). Then the BODIPY solution (50 μL) was added to the carbon suspension. The suspension was then agitated using a Dremel engraver (model 290-01) as a vortex device for 4 h. The particles were subsequently washed with water and dichloromethane to remove the buffer salt and unattached BODIPY. The BODIPY attached carbon particles were dried in vacuum.

6.2.3 Measurement of Fluorescence Quantum Yield

To determine the fluorescence quantum yield, Q, the integrated fluorescence intensity was plotted against the absorption at the excitation wavelength. The absorption intensity was measured using a CARY 5000 UV-Vis-NIR spectrometer (Varian). The fluorescence was measured using a CARY Eclipse Fluorescence spectrophotometer (Varian). The excitation wavelength used here was 480 nm. Although the fluorescence intensive is higher at 485 nm excitation wavelength, the left tail of the emission peak is blocked by the intensive excitation light because the wavelengths of the emission and excitation are too close. In order to measure the quantum yield more accurately, 480 nm was used as the excitation wavelength. The slope, m, was obtained from the linear regression of the plot. A representative plot of Hypercarb is shown in Fig. 6.4. For conventional fluorescent dyes, the straight line goes through the origin. Unlike those dyes, the plot of the carbon particles does not go through the origin. The reason may be that the carbon can also absorb light. But carbon may not transfer energy to the fluorophore. Therefore, the intercept may due

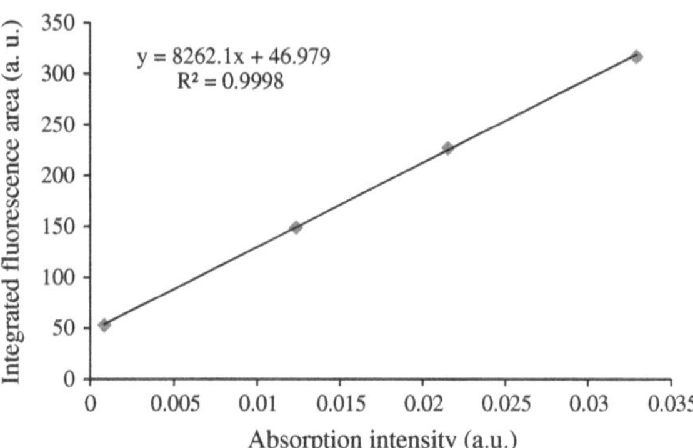

Fig. 6.4 Plot of integrated fluorescence area versus the absorption intensity of BODIPY modified Hypercarb

to the absorption from carbon. The fluorescence quantum yield was measured using the following equation:

$$Q = Q_R(m/m_R)(n^2/{n_R}^2), \tag{6.2}$$

where n is the refractive index of the solvent and the subscript R refers to the reference sample. Rhodamine B in water was used as the reference here. Its quantum yield is 31 % [16]. Water is also used as the solvent for the fluorescent carbon particles. From the slope of the plot of integrated fluorescent intensity versus absorption, the fluorescence quantum yield of BODIPY modified Hypercarb and activated carbon was determined.

6.3 Results and Discussion

Water insoluble carbon particles were prepared by chemically attaching fluorescent dyes onto the carbon surface. To prevent quenching from carbon, a spacer was incorporated between carbon and the fluorescent dyes. In this work the spacer contains two portions. One is from BODIPY that has an 11-atom side chain. The other one is from AHAMTES that also has an 11-atom side chain. The total spacer between the carbon surface and BODIPY fluorophore is a 22-atom chain. For preparation, the carbon particles were pretreated with potassium permanganate for 24 h at room temperature in order to generate hydroxyl groups on the surface [13].

The hydroxylated particles were then reacted with AHAMTES. Initially, this reaction was conducted in dry toluene under nitrogen atmosphere [13]. The reaction was set up in a glove bag purged with nitrogen. After the reaction, the particles were washed with toluene and methanol. SEM image (Fig. 6.5) showed that the particles agglomerated due to the polymerization of the silane-coupling reagent, AHAMTES.

Two other methods have been used to prepare the silanized carbon particles. In the first method, the reaction was conducted in the vapor phase [14]. The silane-

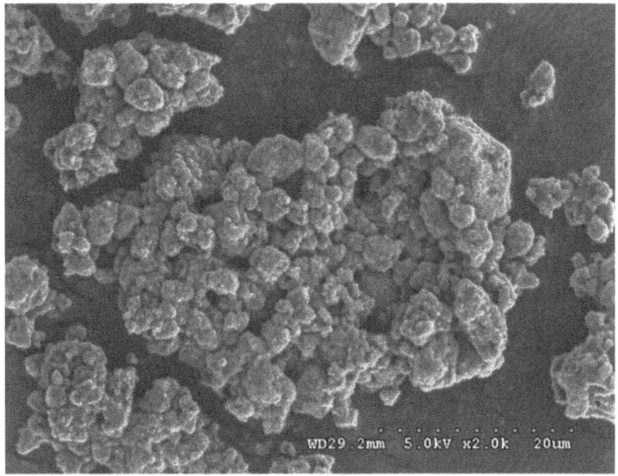

Fig. 6.5 SEM image of the silanized carbon black when the reaction was performed in toluene

coupling reagent was placed in a large flask and the carbon particles were placed in a small container that was placed in the large flask. The contact between the liquid silane-coupling reagent and carbon particles was avoided carefully. The vapor of the silane-coupling reagent was generated by heating the reaction system to a higher temperature. The vapor was transferred by diffusion and reacted with the carbon particles. Under this reaction condition, the polymerization of the silane-coupling reagent on the carbon particles was effectively avoided because the polymerized silane reagent cannot vaporize due to the large molecular weight. In the second method, the reaction was conducted in the solution phase using ethanol as the solvent and the reaction was refluxed for 3 days [15]. Ethanol was used because it can prevent the AHAMTES from polymerization. One of the reaction products of the polymerization is ethanol. In the presence of large amounts of ethanol, the polymerization of the silane reagents was inhibited. Both methods have demonstrated their superior application for the preparation of carbon particles without agglomeration. Solution phase condition was chosen here because the reaction is less sensitive to the water concentration in the system. The surface attachment of the silane coupling reagent was confirmed by the XPS analysis. Two representative XPS spectra from the silanized Hypercarb particles are shown in Fig. 6.6.

BODIPY was then attached to the amine group at the end of AHAMTES. The reaction can be done at room temperature. Because the particles are insoluble the reaction suspension needs to be shaken vigorously. Similar to other fluorescent dyes, BODIPY is photosensitive and can undergo photobleaching if exposed to the light at its absorption wavelength. Therefore, the reaction was performed in a dark room or the reaction vial was protected by aluminum foil from light.

In this work, we measured the fluorescence emission of the BODIPY modified Hypercarb and activated carbon particles in water. The emission of the two types of the particles is at the same wavelength, 510 nm. The reported fluorescence emission of the BODIPY is at 508 nm in aqueous solution [17]. It is very close to

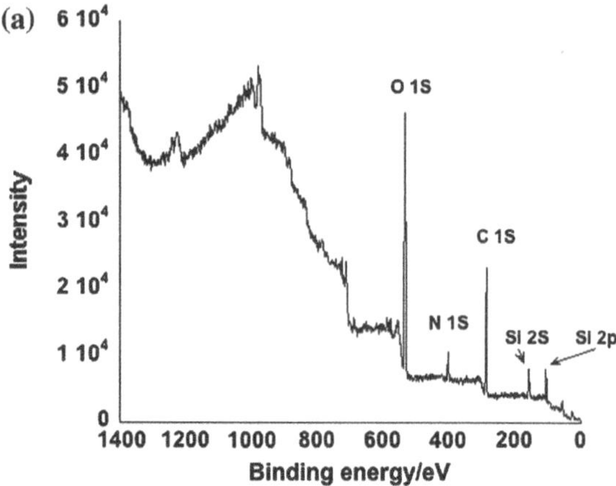

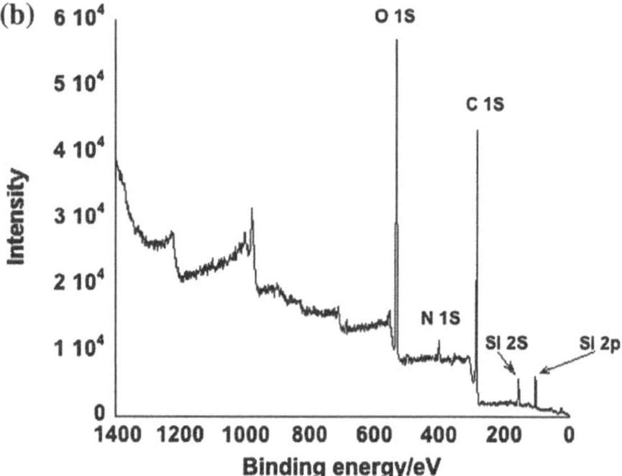

Fig. 6.6 **a** The XPS spectra of the silanized Hypercarb particles using vapor phase reaction and **b** solution phase reaction

the fluorescence wavelength obtained in this work. The result indicates that the emission is from the BODIPY fluorophore.

The emission wavelength of the fluorescent carbon nanoparticles that have been reported shows a dependence on the excitation wavelength [18]. The reason is that the emission is due to the presence of the surface energy traps. However, the photoluminescence of BODIPY modified carbon is due to the attachment of the fluorophore. Figure 6.7 shows that the emission wavelength of the BODIPY modified carbon particles is independent of the excitation wavelength. It is a strong evidence that the fluorescence of the carbon particles is from the BODIPY fluorophore. When the excitation wavelength increases the intensity of the emission also increases.

The fluorescence quantum yield of BODIPY modified Hypercarb was determined to be 0.26 % and the fluorescence quantum yield of BODIPY modified activated carbon was 0.48 %. The fluorescence quantum yield of BODIPY modified carbon particles is low. There are two possible reasons. (1) The quantum yield

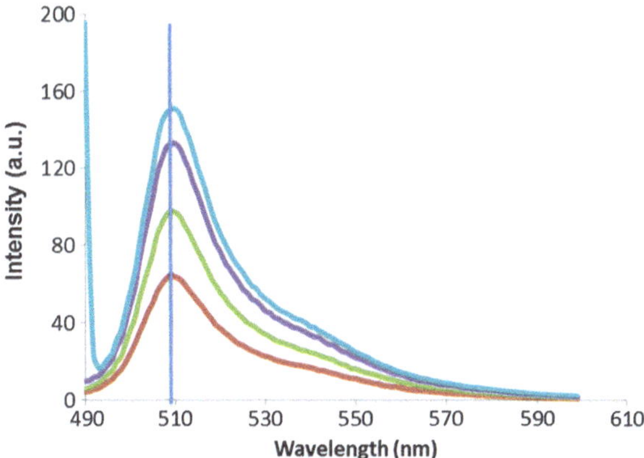

Fig. 6.7 Fluorescence spectra of BODIPY modified activated carbon using different excitation wavelengths: — 460 nm, — 470 nm, — 480 nm, — 485 nm

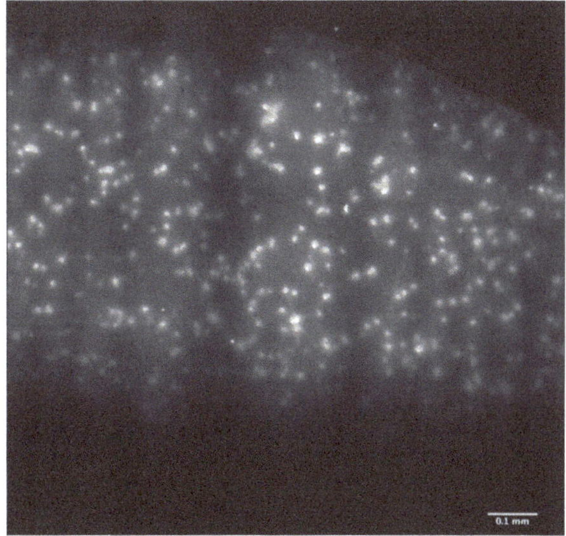

Fig. 6.8 The *photograph* of the fluorescent carbon particles prepared from BODIPY modified Hypercarb

is determined by the slope of the area of the emission peak versus the absorption intensity. Carbon has strong absorption in a wide range of wavelength [19]. However, carbon may not transfer the energy to the fluorophore. In that case, the fluorescence quantum yield from the absorption of carbon is zero. (2) Although a spacer is used to separate the fluorophore and carbon, the chain may bend in solution and the distance between carbon and BODIPY would be shorter. As a result, the fluorescence may still be quenched by carbon. Ideally, the fluorescence quantum yield of Hypercarb and activated carbon should be same because the fluorescence is from BODIPY. The difference of the measured fluorescence quantum yield may be because the absorption cannot be measured accurately. The measured absorption includes the scattering of the light from the carbon particles in addition to the absorption.

Although the quantum yield of the BODIPY modified carbon particles was less than 1 %, it can be used for the tracers for imaging the movement of the particles. A photograph of the fluorescent carbon particles prepared from Hypercarb is shown in Fig. 6.8. The photograph was taken by Professor Minami Yoda, Professor of Mechanical Engineering, et al. from Georgia Institute of Technology. The bright spots are fluorescent carbon particles.

6.4 Conclusion

Water insoluble fluorescent carbon particles were prepared by chemically attaching a BODIPY fluorophore to the carbon surface. A spacer of 22-atoms was placed in between to avoid the fluorescence quenching from carbon. Compared to other fluorescent carbon, the particles prepared by this method exhibit much lower solubility in water. Moreover, the fluorescent carbon particles with a variety of sizes can be easily prepared.

References

1. Baker SN, Baker GA (2010) Angew Chem Int Edit 49:6726
2. Wereley ST, Meinhart CD (2010) Annu Rev Fluid Mech 42:557
3. Sun YP, Zhou B, Lin Y, Wang W, Fernando KAS, Pathak P, Meziani MJ, Harruff BA, Wang X, Luo PG, Yang H, Kose ME, Chen B, Veca LM, Xie SY (2006) J Am Chem Soc 128:7756
4. Xu X, Ray R, Gu Y, Ploehn HJ, Gearheart L, Raker K, Scrivens WA (2004) J Am Chem Soc 126:12736
5. Peng H, Travas-Sejdic J (2009) Chem Mater 21:5563
6. Ray SC, Saha A, Jana NR, Sarkar R (2009) J Phy Chem 113:18546
7. Satishkumar BC, Brown LO, Gao Y, Wang C, Wang H, Doorn SK (2007) Nat Nanotechnol 2:560
8. Lakowicz JR (1999) Principles of fluorescence spectroscopy. Kluwer Academic/Plenum Publishers, New York
9. Matt RM, Voss EW (1977) Immunochemistry 14:533
10. Caminati G, Peroni E, Papini AM, Baglioni P (2004) Progr Colloid Polym Sci 126:163
11. Baker JG, Middleton R, Adams L, May LT, Briddon SJ, Kellam B, Hill S (2010) J Br Pharmacol 159:772
12. Vu TT, Badre S, Dumas-Verdes C, Vachon J, Julien C, Audebert P, Senotrusova EY, Schmidt EY, Trofimov BA, Pansu RB, Clavier G, Meallet-Renault R (2009) J Phys Chem C 113:11844
13. Kim M, Hong J, Hong CK, Shim SE (2009) Synth Met 159:62
14. Zhu M, Lerum MZ, Chen W (2012) Langmuir 28:416
15. Ansell M, Cogan EB, Page CJ (2000) Langmuir 16:1172
16. Magde D, Rojas GE, Seybold P (1999) Photochem Photobiol 70:737
17. Kang S, Jin S, Kim D (2007) Anal Biochem 360:1
18. Zhang Y, Goncalves H, Esteves da Silva JCG, Geddes CD (2011) Chem Commun 47:5313
19. Taft EA, Philipp HR (1965) Phys Rev 138:197

Chapter 7
Future Work

Abstract This chapter briefly describes future work based on this dissertation.

7.1 UTLC Future Work

Electrospun UTLC with PAN [1], glassy carbon [2] and PVA [3] has been reported and described in Chap. 2. The stationary phases with different functionalities provide excellent separations for steroids, amino acids, etc. Sensitive detection is an advantage of UTLC analysis. Only small amount of sample is required for the analysis and detection. However, due to the small thickness of the stationary phase the loading capability is limited. Sample spotting has to be performed very carefully in order to prevent overloading the plate. To solve this problem, electrospun porous nanofibers could be used. As described in Chap. 5, porous PAN with different porosity can be prepared when the amount of sodium bicarbonate salts is controlled [4]. The porous nanofibers have higher surface area. As a result, the loading capacity can be improved. Therefore, porous nanofibers with different porosity, such as PAN, can be prepared for UTLC application. The loading capacity will be studied and compared.

7.2 Homogeneous Carbon Future Work

Homogenous edge-plane carbon as a stationary phase has demonstrated its potential in the separation of polar analytes in Chap. 3. In Chap. 4, homogenous basal-plane with different adsorption selectivity is described. Application of homogenous basal-plane carbon in HPLC separation is of great interest in the separation of nonpolar analytes and isomers. In order to prepare the basal-plane carbon packing for HPLC, a similar strategy described in Chap. 4 can be used. The silica gel particles or alumina particles can be coated with silver. Then AR MP can be coated on to the silver coated particles. In that way a face-on anchoring state is expected.

T. Lu, *Nanomaterials for Liquid Chromatography and Laser Desorption/Ionization Mass Spectrometry*, Springer Theses,
DOI: 10.1007/978-3-319-07749-9_7,

Because the basal-plane carbon is composed of large π conjugated graphite layers. The basal-plane surface can have dispersive interactions with analytes. When separating aromatic analytes, they can have strong π-π interactions with the basal-plane surface. A recent paper has reported that using graphene as the stationary phase for open tubular capillary LC [5]. When the graphene was used the aromatic analytes were poorly separated. The reason may be the strong π-π stacking between the graphene and aromatic rings. When they used oxidized graphene, the separation was improved. The organic functional groups on the graphene oxide decrease the π-π stacking between the graphene and aromatic rings. Therefore, if the basal-plane carbon shows similar separation results to the graphene, oxidizing the basal-plane surface may solve the problem. On the other hand, homogenous edge-plane and basal-plane carbon can be used as stationary phases for open tubular LC. Compared to packed column, open tubular LC columns provide much faster separation and generate much lower back pressure. Low temperature glassy carbon as a stationary phase has been used for open tubular supercritical fluidic LC [6]. The similar preparation method can be used to prepare homogenous edge-plane and basal-plane carbon coated open tubular capillary columns. AR MP can be coated to the fused silica capillary and after pyrolysis homogenous edge-plane carbon can be formed. In order to prepare the basal-plane carbon coated silica capillary, silver coating is required. Another way is to use the homogenous carbon nanorods that are prepared for SPE. The carbon nanorods can be attached onto the silica surface by diazonium salt chemistry [7]. The reaction is shown in Fig. 7.1. The surface of the fused silica capillary is modified with a silane reagent containing aminophenol group, such as p-aminophenylmethoxysilane. The reduction of the amino group on the benzene ring to diazonium salt and the attachment of benzene ring to the carbon surface can be achieved in one step [7].

One drawback of the open tubular LC is the low separation efficiency that is limited by the slow diffusivity in liquid phase. Capillaries with very diameters, <10 μm, are required [8]. Injector with very small volume and sensitive detector are needed. In addition, the modification of the inner wall of capillaries becomes more difficult when the diameter of the tubing decreases. Enhanced-fluidic LC (EFLC) provides much faster diffusion using carbon dioxide mixed with the mobile phase [9]. Therefore, using open tubular capillaries, 50–100 μm id, with homogenous carbon stationary phases for EFLC should be a more promising direction.

7.3 Laser Desorption/Ionization MS Future Work

Chapter 5 discusses the application of electrospun carbon and polymer in laser desorption/ionization MS. The work is mainly focused on the SALDI analysis of synthetic polymers and small molecules. Other SALDI methods are also a powerful tool for the analysis of biomolecules in addition to synthetic polymers. Therefore, using electrospun nanofibrous substrates for the SALDI analysis of biomolecules, proteins and peptides, needs to be studied.

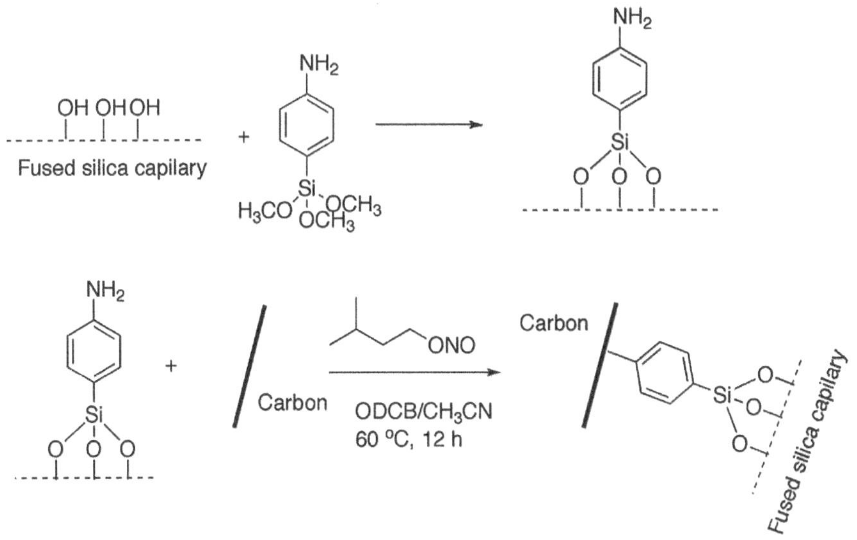

Fig. 7.1 Synthetic scheme for attachment of carbon nanorods onto the fused silica capillary using diazomium salt chemistry. *ODCB* 1,2-dichlorobenzene

On the other hand, SELDI has been used for the analysis of proteins in complex sample solutions [10]. As discussed in Chap. 1, the first step in SELDI is to purify the sample by washing the sample spot on the plate. Because the analyte has strong interaction with the SELDI substrates, usually polymers, the hydrophilic impurities can be removed during the washing step. Then the organic matrix is added and the analyte remaining on the plate is analyzed using the conventional MALDI method. The difference between SELDI and SALDI is that organic matrixes are required in SELDI analysis. Electrospun carbon and polymeric nanofibers have great potential as substrates for SELDI. One reason is that carbon and PAN and SU-8 polymers can interact with proteins, which is required for the purification step. Another reason is that the nanostructured substrates have high surface area, which may increase the amount of the protein that is adsorbed on the substrates. The limit of detection may be improved. Finally, because carbon and polymeric substrates can be used as SALDI substrates as well, organic matrix may not be required for the SELDI analysis. Without using organic matrix, the sample preparation is more simple, sample distribution is more homogenous, and searching of "sweet spot" is less demanding.

References

1. Clark J, Olesik SV (2009) Anal Chem 81:4121
2. Clark J, Olesik SV (2010) J Chromatogr A 1217:4655
3. Lu T, Olesik SV (2013) J Chromatogr B 912:98
4. Ma G, Yang D, Nie J (2009) Polym Adv Technol 20:147–150
5. Qu Q, Gu C, Hu X (2012) Anal Chem 84:8880
6. Engel TM, Olesik SV, Callstrom MR, Diener M (1993) Anal Chem 65:3691
7. Bahr JL, Tour JM (2001) Chem Mater 13:3823
8. Pfeffer WD, Yeung ES (1990) Anal Chem 62:2178
9. Cui Y, Olesik SV (1995) J Chromatogr A 691:151
10. Zaluzec EJ, Gage DA, Allison J, Watson JT (1994) J Am Soc Mass Spectrom 5:230

VITA

May, 2005	B.S. Chemistry, Peking University
May, 2008	M.S. Chemistry, North Dakota State University
August, 2013	Ph.D. Analytical Chemistry, The Ohio State University

T. Lu, *Nanomaterials for Liquid Chromatography and Laser Desorption/Ionization Mass Spectrometry*, Springer Theses,
DOI: 10.1007/978-3-319-07749-9,

Zeitfracht Medien GmbH
Ferdinand-Jühlke-Straße 7
99095 Erfurt, Deutschland
produktsicherheit@kolibri360.de